Practical Booklet of Linux 1

Baldomero Sánchez Pérez

First edition
Date of publication: 19/02/2018
ISBN: 978-0-244-36963-7
Edited by: Lulu.com

This book is dedicated to all people who wish to learn Linux and want to be trained on a practical level in the Linux operating system to get a degree in the medium-level training cycle as a Technician in Microcomputer Systems and Networks or Higher Degree Computer Systems (DAW, DAM), or simply acquire the basic administrator information in Linux.

To appreciate that perseverance and longing for learning as enthusiastic that I place in myself and in my sister, the one who saw us grow and persevered that we should study, my dear mother.

"You need to study other people's work. Their approaches to solving problems and the tools they use give you a new way of looking at your own work"

Gary Kildall

Source/Notes:
Programmers at Work (1986)

INDEX

PREFACE

This book is called a practical notebook, because it includes the concepts of didactic programming, work units that form the modular blocks covered (Virtual Machines, Theory of Linux operating systems and the Linux and Android Operating System). The units of work organized in a sequence of practices have been collected.

Each practice is organized, with an objective to achieve, the description of the knowledge for the development of the practice, the necessary requirements for the installation, management and the steps that must be followed for its development, (brief or extensive). The practices collect illustrations or results obtained, based on a specific version made on a Virtual Box machine.

The practices contain clarifying complements, of their realization or previous knowledge, related to their development at a practical or theoretical level. They are accompanied by vignettes or clarification of their development, reflected in different colors: Blue explanatory note, Light green prerequisites, Orange important notes or precautions.

The methodology used in the development of the practices is a Constructivist methodology, which starts from "From the concrete to the Abstract", "From the known to the unknown", "From the general to the particular". The initial "conductivist" learning of basic structures is intended, so that the student must be able to learn and deduce from their own experiences guided by the teacher (teacher) in the essential. Although it can be used in distance training, as in teaching self-taught people.

The methodological aspects that are intended to be applied in the programming are based on the idea, that the student considers himself an active part of the teaching activity, promoting self-learning and improving the knowledge itself.

The aim is to involve the student in the process of assimilating new concepts and acquiring skills, to prepare the student as an active member of today's society.

All software used is GPL or Copyright © of registered trademarks in which no change has been made, example VirtualBox of Oracle © and other LINUX versions are GNU General Public License (GPL).

WORK UNIT I: Introduction to the storage of Desktop Operating Systems.

PRACTICE 1: Access a NTFS file system from Linux.

PRACTICE 2: Access and repair data from a Windows file system from Linux Ubuntu.

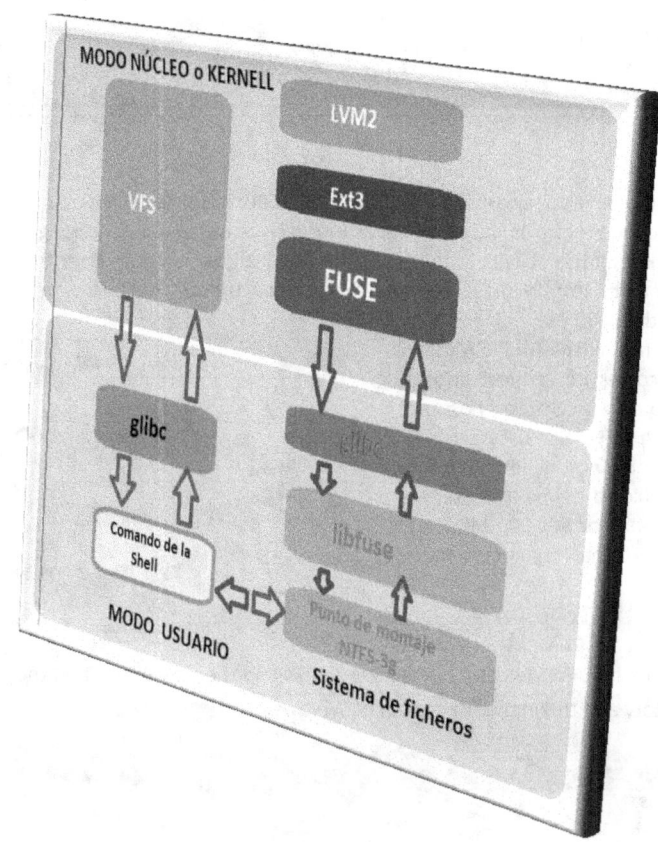

Commands
/etc/fstab, /etc/mtab
fdisk, gdisk, cgdisk,
cfdisk, mount,
umount, fsck, bklib,
tune2fs, uuidgen,
edquota, quota,
quotacheck, quotaon,
quotaoff, repquota,
warnquota

Contents
- Partitioning units.
- Types of partitions.
- The file system.
- Types of file systems.

PRACTICE 1: Access a NTFS file system from Linux.

DESCRIPTION:

Windows NT was designed from the beginning to be a network operating system and multitasking that definitively broke any link with its MS-DOS ancestors, for which a new file system was designed based on a radically new design (it is not therefore of a new bodywork of the previous FAT).

The resulting system, called NTFS ("New Technology File System") is a very robust system that allows compression of files one by one; a very developed protocol for use authorization and file attributes; transaction system based on transactions; RAID support; possibility of joining the capacities of two units in a single volume ("Disk striping") and many other improvements, such as the ability to note bad clusters ("Hot fixing") in run-time.

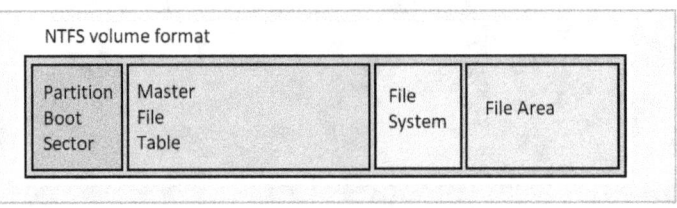

Its penultimate version, the so-called NTFS 5, incorporated in Windows 2000, has some other advanced features, such as file encryption support built into the OS itself; properties of files based on persistent user identifiers (it is no longer necessary to identify files by their endings), and unique identification of all the objects in the file system that allows, among other things, that a file can occupy different volumes (files multivolume). Although naturally these benefits charge their tribute. NTFS uses very large meta-structure, so it is not recommended for volumes of less than 400 GB.

The central structure of this system is the MFT ("Master File Table"), from which several copies of its most critical part are kept in order to protect it against possible corruptions. Like FAT16 and FAT32, NTFS also uses clusters as a storage unit, although these do not depend on the volume of the partition. It is possible to define a cluster of 512 bytes (1 sector) in a 5 MB partition or 500,000 MB. This capacity reduces both internal and external fragmentation.

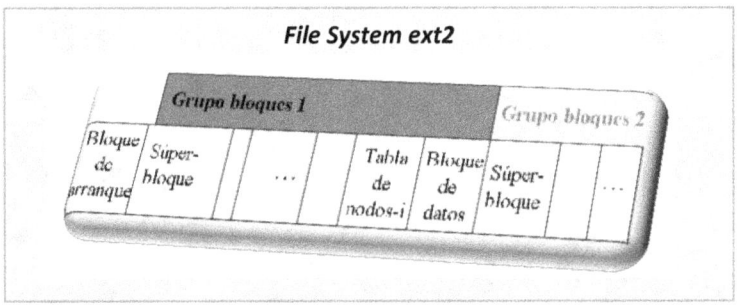

Using NTFS from Linux (ext partitions)

Boot from a Linux, it is installed in RAM memory. The structure of the file system is created in RAM. Access to view all the disks and their partitions and they are mounted in the boot system. Once mounted access to the directory and this in the file system of Widows.

STEP 1: Boot from the Linux ISO.

Visualize the disks and partitions.

```
root@slackware:/# fdisk -l
Disk identifier: 0x7d6d7d6d
```

Device Boot	Start	End	Blocks Id	System
/dev/sdd1	2048	46135295	23066624 7	HPFS/NTFS/extFAT

```
Disk /dev/sde: 23.6 GB, 23622320128 bytes
255 heads, 63 sectors/track, 2871 cylinders, total 46137344 sectors
Units = sectors of 1 * 512 = 512 bytes
Sector size (logical/physical): 512 bytes / 512 bytes
I/O size (minimum/optimal): 512 bytes / 512 bytes
Disk identifier: 0x1169e640
```

Device Boot	Start	End	Blocks	Id	System
/dev/sde1	63	4096574	2048256	6	FAT16
/dev/sde2	4096575	26619704	11261565	f	WP5 Ext' d (LBA)
/dev/sde5	4096638	6152894	1028128+	6	FAT16
/dev/sde6	6152958	8209214	1028128+	6	FAT16
/dev/sde7	8209278	10265534	1028128+	6	FAT16
/dev/sde8	10265598	12321854	1028128+	6	FAT16
/dev/sde9	12321918	14378174	1028128+	6	FAT16
/dev/sde10	14378238	16434494	1028128+	6	FAT16
/dev/sde11	16434558	19294064	1429753+	7	HTPFS/NTFS/exFAT
/dev/sde12	20547198	22603454	1028128+	6	FAT16
/dev/sde13	22603518	26619704	2008093+	6	FAT16

```
root@slackware:/# _
```

STEP 2: Mount a file system.

     mount

a) Create a directory.

mkdir	create a directory.
cd	Access directory.
ls -l	Visualize dictory (dir).
cd mnt	Access directory mnt.
ls −l	View the contents of the mnt directory.
mkdir win7	Create the win7 directory inside mnt.
mkdir winxp	Create the winxp directory inside mnt at the same level as win7..

b) Mount Windows file systems, at a Linux mount point.

     /dev/sdc1   /mnt/winxp

     /dev/sde1   /mnt/win7

c) Identified the partitions and the directories of the assembly point, the assembly and

     mount /dev/sdc1 /mnt/winxp

     cd winxp

     ls -l

     cd /mnt

     ls -l

     mount /dev/sde1 /mnt/win7

     cd /mnt/win7

     ls −l

> To assemble, I must be outside the directory that will be the assembly point. Mount Windows 7, current directory / mnt:
>   # mount /dev/sde11   /mnt/win7
>   # cd win7
>   # ls -l

d) Reassemble a file system from the command line, without booting the Operating System.

     mount -o remount,rw /

STEP 3: Disassemble a mounting point.

     umount

a) Help.

     umount --help

b) View mount points that are currently mounted.

     mount

c) Remove the mounting point /mnt/winxp.

     mount

     umount /mnt/winxp

     cd winxp

     ls -l

First we visualized the assembly points, we removed the assembly point and later we accessed the directory where the assembly point had been made and we visualized the contents of the directory.

TYPE	DESCRIPTION
auto	try to discover the file system automatically
iso9660	File system: CD and DVD
ext2	GNU / Linux native file system
ext3	GNU / Linux native file system
ext4	GNU / Linux native file system
reiserfs	GNU / Linux native file system
msdos	sistemblinka of FAT files
fat	FAT16 file system
vfat	FAT32 file system
ntfs	NTFS file system in reading mode
ntfs-3g	NTFS file system in read and write mode
smbfs	SAMBA server file system
nfs	NFS network file system of GNU / Linux
hfs	Apple Macintosh file system
hfsplus	Apple Macintosh file system
ncpfs	Novell NetWare file system.

STEP 4: Automatic mount point of file systems when starting.

The file /etc/fstab Contains a line with the assembly specifications of each file system on which we normally work: the file system in which we have the Linux directories, the /proc, the partition two, the CDROM, and the Floppy.

The /etc/fstab file works as follows.

a)   We start with an example of content from / etc / fstab:

# \<device\>	\<mountpoint\>	\<filesystemtype\>	\<options\>	\<dump\>	\<fsckorder\>
/dev/hda2	/	ext2	defaults	1	1
/dev/hda3	/usr	ext2	defaults	1	2
/dev/sda1	/home	ext2	defaults	1	2
/dev/hdb	/mnt/cdrom	iso9660	user,noexec,nodev,nosuid,ro,noauto	0	0
/dev/fd0	/mnt/floppy	vfat	user,noexec,nodev,nosuid,rw,noauto	0	0
none	/	proc	proc defaults	0	0
/dev/hda4	swap	swap	defaults	0	0
/dev/hda1	/mnt/dos	vfat	exec,dev,suid,rw,auto	0	0

With the information of the mount point devices of the file systems, the operating system would read the following file /dev/fstab and execute the file systems and mount points with their options and values.

- The /dev/hda1 partition would be mounted in the subdirectory /mnt/two
- The /dev/hda2 partition would be mounted in the subdirectory /
- The /dev/hda3 partition would be mounted in the /usr subdirectory
- The /dev/hda4 partition would be mounted in the subdirectory as swap
- The /dev/sda1 partition would be mounted in the /home subdirectory
- Proc would be mounted in the subdirectory /proc
- The system would have information on how to mount a floppy disk /dev/fd0 and a CD-ROM /dev/hdb, although it does not mount them automatically when booting because you have defined the noauto option.

b)   The parameters used in /etc/fstab:

In the device column the device / partition to be mounted is indicated, in the mount point the directory by which we are going to access the file system is indicated. The file system type column (filesystemtype) indicates the file system that will be used on the device.

<device>   <mountpoint> <filesystemtype> <options>   <dump> <fsckorder>

<device>: in this field the device or partition where the filesystem is located is indicated, the UUID is currently displayed.

DEVICE IDENTIFICATION	DESCRIPTION
/dev/sda1	The name of the device. (Standard)
LABEL=SISTEMA-OPERATIVO	The tag is assigned as the file system identifier.
UUID= cbf5fdfb-ff17-4b38-9ee6-57492ba5482	UUID = is used followed by the UUID without quotes.
PARTLABEL=SISTEMA-ARRANQUE	The label is used as identification of the GPT label file system.
PARTUUID= cbf5fdfb-ff17-4b38-9ee6-57492ba5482d	The own UUID partition of GPT boot loaders UUID is identified, use the UUID values without quotes.
//host/comparte	The name of the server and the share are identified.
/dev/sdg1	Load external devices.
UUID=47FA-4071	File space indicates the path of the files.

Ej.:
```
/dev/sda2              none        swap       defaults           0     0
/dev/sda3              /home       ext4       defaults,noatime   0     2
LABEL=DATA         /home ext4    rw,relatime,discard,data=ordered   0     2
UUID=CBB6-24F2                         /boot vfat
rw,relatime,fmask=0022,dmask=0022,codepage=437,iocharset=iso8859-1,shortname=mixed,errors=remount-ro 0
2
PARTUUID=039b6c1c-7553-4455-9537-1befbc9fbc5b none  swap    defaults    0     0
PARTLABEL=HOME                        /home ext4   rw,relatime,discard,data=ordered 0     2
UUID=47FA-4071      /home/username/Camera\040Pictures    vfat  defaults,noatime       0  0
/dev/sda7          /media/100\040GB\040(Storage)       ext4  defaults,noatime,user  0  2
/dev/sdg1          /media/backup    jfs    defaults,nofail,x-systemd.device-timeout=1    0  2
//192.168.2.100/comparte    /net/share cifs  noauto,nofail,x-systemd.automount,x-
systemd.requires=network-online.target,x-systemd.device-
timeout=10,workgroup=workgroup,credentials=/foo/credentials          0 0
UUID=47FA-4071      /home/username/Camera\040Pictures    vfat  defaults,noatime       0  0
/dev/sda7          /media/100\040GB\040(Storage)       ext4  defaults,noatime,user  0  2
```

<mountpoint>: here goes the mount point for the specified device.

<filesystemtype>: the type of file system. It can take several values, among which stand out: ext2, ext3, ext4, iso9660, nfs, ntfs, reiserfs, smbfs, swap, vfat, xfs.

<option>: in this column are the options for mounting the filesystem. There are many and the most common are mentioned below. For a more complete list, you can read the manual of the mount command and that of the nfs (for the specific parameters of nfs).

NOTE: As the spaces are used in fstab to delimit fields, if any field (PARTLABEL, LABEL or the mount point) contains spaces, these spaces must be replaced by escape characters \ followed by the octal code of 0 digits 040

<options>	<dump>	<fsckorder>
defaults	1	1
defaults	1	2
defaults	1	2
user,noexec,nodev,nosuid,ro,noauto	0	0
user,noexec,nodev,nosuid,rw,noauto	0	0
proc defaults	0	0
defaults	0	0
exec,dev,suid,rw,auto	0	0

Field options <options>	
<options>	DESCRIPTION
user, nouser	allows / does not allow an ordinary user to mount the file system.
suid, nosuid	It allows / does not allow having files with the defined user bit.
auto/noauto	Indicates yes / no mount when we do mount -a.
defaults	Apply the options rw, suid, dev, exec, auto, nouser, async.
exec/noexec	It allows / does not allow the execution of binaries.
ro, rw	Mount read only, read-write.
sync/async	All I/O accesses to the file system will be made in synchronous / asynchronous mode.

<dump>: this column tells the dump utility whether or not to backup the filesystem. You can take two values: 0 and 1. With 0 it is indicated that you should not backupear, with 1 that is done. Logically, it depends on having installed and configured dump, so in most cases this field is 0.

<fsckorder>: In this case it is an indication for the fsck (command that checks the filesystem) and is again defined with a numerical value. The possibilities are 0, 1 and 2. The 0 indicates that the filesystem should not be checked, while the 1 and 2 tell fsck to check it. The difference is that 1 represents a higher priority than 2, so it must be used for the root system and 2 for the rest of the file systems.

c)   The different types of file systems that are supported in /etc/fstab.

STEP 5: Display the UUID of the devices.

blkid

a) By default, it visualizes the filesystems mounted with their UUIDs.

blkid

```
[  408.636167] print_req_error: I/O error, dev fd0, sector 0
/dev/sda1: UUID="bnx2oE-UtGp-oGAW-euKJ-E3uM-fN27-8KsAuM" TYPE="LVM2_member" PARTUUID="937101c5-01"
/dev/mapper/ubuntusvr1710--vg-root: UUID="1c78bb35-0c3a-4b5c-9e3b-60669ea69333" TYPE="ext4"
/dev/mapper/ubuntusvr1710--vg-swap_1: UUID="50e6d52e-14ab-48ce-be11-3e60e29329cb" TYPE="swap"
```

b) Display all file systems, in a single column, as if it were an enumeration.

bklid –k
linux_raid_member
dff_raid_member
isw_raid_member
....

c) Garbage collection. The devices that are in memory appear again.

blkid –c /dev/null

```
[  408.636167] print_req_error: I/O error, dev fd0, sector 0
/dev/sda1: UUID="bnx2oE-UtGp-oGAW-euKJ-E3uM-fN27-8KsAuM" TYPE="LVM2_member" PARTUUID="937101c5-01"
/dev/mapper/ubuntusvr1710--vg-root: UUID="1c78bb35-0c3a-4b5c-9e3b-60669ea69333" TYPE="ext4"
/dev/mapper/ubuntusvr1710--vg-swap_1: UUID="50e6d52e-14ab-48ce-be11-3e60e29329cb" TYPE="swap"
```

d) Do not display the non-printable character. In this case the same thing as case c).

blkid -d

e) Formats to visualize the output (-o| --output format): full, value, list, device, udev, export.

blkid -o value
blkid -o list

```
root@ubuntusvr1710:~# blkid  -o list
device                         fs_type        label         mount point              UUID
----------------------------------------------------------------------------------------------------
[ 2736.416114] print_req_error: I/O error, dev fd0, sector 0
/dev/sda1                      LVM2_member                  (in use)                 bnx2oE-UtGp-oGAW-euKJ-E3uM-fN27-8KsAu
M
/dev/mapper/ubuntusvr1710--vg-root    ext4                  /                        1c78bb35-0c3a-4b5c-9e3b-60669ea69333
/dev/mapper/ubuntusvr1710--vg-swap_1  swap                  [SWAP]                   50e6d52e-14ab-48ce-be11-3e60e29329cb
```

f) Test all systems except RAIDs.

blkid --- probe –usages /dev/sda1

```
/dev/sda1: PART_ENTRY_SCHEME="dos" PART_ENTRY_UUID="937101c5-01" PART_ENTRY_TYPE="0x8e" PART_ENTRY_FLAGS="0x80" PART_ENTRY_NUMBER="1" PART_ENTRY
_OFFSET="2048" PART_ENTRY_SIZE="266334208" PART_ENTRY_DISK="8:0"
```

g) Another way to visualize the UUIDs of the devices mounted in LINUX, are located the links in the directory / dev / disk / by-uuid.

ls -l /dev/disk/by-uuid

```
lrwxrwxrwx 1 root root 10 nov 26 08:37 1c78bb35-0c3a-4b5c-9e3b-60669ea69333 -> ../../dm-0
lrwxrwxrwx 1 root root 10 nov 26 08:37 50e6d52e-14ab-48ce-be11-3e60e29329cb -> ../../dm-1
```

h) Consult the UUID of a specific partition.

tune2fs -l /dev/sda1 | grep UUID
blkid /dev/sda1

STEP 6: Generate a UUID, to then assign it to a device.

uuidgen

a) Generates by default a new UUID, randomly.

uuidgen
cbf5fdfb-ff17-4b38-9ee6-57492ba5482d

STEP 7: Assign a new UUID to a device.

tune2fs

a) The UUID generated by uuidgen is used and assigned to the new device.

tune2fs – U 'cbf5fdfb-ff17-4b38-9ee6-57492ba5482d' /dev/sda2

STEP 8: We check the filesystem system label.

blkid

a) Find out the file system by consulting all the information of a file system by its UUID

blkid /dev/sda1

```
/dev/sda1: UUID="bnx2oE-UtGp-oGAW-euKJ-E3uM-fN27-8KsAuM" TYPE="LVM2_member" PARTUUID="937101c5-01"
```

b) Display the label of a file system with tune2fs.

tune2fs -l /dev/sda1 | grep "Filesystem volume name"

STEP 9: Set the system label filesystem.
        tune2fs
a)   Change the label of a file system.
        tune2fs -L "SISTEMA-OPERATIVO" /dev/sda1

STEP 10: Contains a list of the mounted file system file /etc/mtab.

   /etc/mtab has a structure very similar to the fstab file, the difference is that **/etc/fstab** is a configuration file that contains the file systems that must be mounted at runtime, as well as its mount points, while mtab, simply lists the file systems mounted at this time.
        cat   /etc/mtab

```
root@ubuntusvr1710:~# cat /etc/mtab
sysfs /sys sysfs rw,nosuid,nodev,noexec,relatime 0 0
proc /proc proc rw,nosuid,nodev,noexec,relatime 0 0
udev /dev devtmpfs rw,nosuid,relatime,size=436664k,nr_inodes=109166,mode=755 0 0
devpts /dev/pts devpts rw,nosuid,noexec,relatime,gid=5,mode=620,ptmxmode=000 0 0
tmpfs /run tmpfs rw,nosuid,noexec,relatime,size=92616k,mode=755 0 0
/dev/mapper/ubuntusvr1710--vg-root / ext4 rw,relatime,errors=remount-ro,data=ordered 0 0
securityfs /sys/kernel/security securityfs rw,nosuid,nodev,noexec,relatime 0 0
tmpfs /dev/shm tmpfs rw,nosuid,nodev 0 0
tmpfs /run/lock tmpfs rw,nosuid,nodev,noexec,relatime,size=5120k 0 0
tmpfs /sys/fs/cgroup tmpfs ro,nosuid,nodev,noexec,mode=755 0 0
cgroup /sys/fs/cgroup/unified cgroup2 rw,nosuid,nodev,noexec,relatime 0 0
cgroup /sys/fs/cgroup/systemd cgroup rw,nosuid,nodev,noexec,relatime,xattr,name=systemd 0 0
pstore /sys/fs/pstore pstore rw,nosuid,nodev,noexec,relatime 0 0
cgroup /sys/fs/cgroup/devices cgroup rw,nosuid,nodev,noexec,relatime,devices 0 0
cgroup /sys/fs/cgroup/pids cgroup rw,nosuid,nodev,noexec,relatime,pids 0 0
cgroup /sys/fs/cgroup/rdma cgroup rw,nosuid,nodev,noexec,relatime,rdma 0 0
cgroup /sys/fs/cgroup/memory cgroup rw,nosuid,nodev,noexec,relatime,memory 0 0
cgroup /sys/fs/cgroup/hugetlb cgroup rw,nosuid,nodev,noexec,relatime,hugetlb 0 0
cgroup /sys/fs/cgroup/net_cls,net_prio cgroup rw,nosuid,nodev,noexec,relatime,net_cls,net_prio 0 0
cgroup /sys/fs/cgroup/freezer cgroup rw,nosuid,nodev,noexec,relatime,freezer 0 0
cgroup /sys/fs/cgroup/blkio cgroup rw,nosuid,nodev,noexec,relatime,blkio 0 0
cgroup /sys/fs/cgroup/cpuset cgroup rw,nosuid,nodev,noexec,relatime,cpuset 0 0
cgroup /sys/fs/cgroup/cpu,cpuacct cgroup rw,nosuid,nodev,noexec,relatime,cpu,cpuacct 0 0
cgroup /sys/fs/cgroup/perf_event cgroup rw,nosuid,nodev,noexec,relatime,perf_event 0 0
systemd-1 /proc/sys/fs/binfmt_misc autofs rw,relatime,fd=24,pgrp=1,timeout=0,minproto=5,maxproto=5,direct,pipe_ino=14747 0 0
hugetlbfs /dev/hugepages hugetlbfs rw,relatime,pagesize=2M 0 0
debugfs /sys/kernel/debug debugfs rw,relatime 0 0
mqueue /dev/mqueue mqueue rw,relatime 0 0
fusectl /sys/fs/fuse/connections fusectl rw,relatime 0 0
configfs /sys/kernel/config configfs rw,relatime 0 0
lxcfs /var/lib/lxcfs fuse.lxcfs rw,nosuid,nodev,relatime,user_id=0,group_id=0,allow_other 0 0
tmpfs /run/user/0 tmpfs rw,nosuid,nodev,relatime,size=92612k,mode=700 0 0
```

PRACTICE 2: File system partitions establishing quotas.

DESCRIPTION:

Types of quota:

> By blocks (blocks) 1 block = 1kb
>
> By inodes (inodes) One inode = 1 file in symbolic links

Limits of use:

> HARD (hard). By blocks or inodes, with absolute limit. The user can not exceed that limit.
>
> SOFT (soft), limit by blocks or inodes, is less than HARD. It can be exceeded by the user.

Applying the fee to users

We must apply the fee per user, although the file system already supports quotas and are enabled, by default no user has established quotas. So to start you will have to manage each user through the command edquota, which will open the text editor that has by default and will show the following:

```
# edquota -u user1
Disk quotas for user user1 (uid 502):
  Filesystem                 blocks      soft      hard     inodes      soft      hard
  /dev/sda3                      56         0         0         14         0         0
```

The columns "blocks" and "inodes" are informative, that is, they indicate the number of blocks or inodes currently used by the user, and those that we can edit are the "soft" and "hard" columns of each case. You can set values by blocks, by inodes or both, remember that the soft limit must be less than hard. If only the hard is set, there will be no previous warnings and the user will no longer be able to save files when the set value is reached. If soft and hard is set, it will warn when the soft limit is exceeded and the grace period will come into play. If the time of thanks is over or the hard is reached (whichever comes first), no more files can be created until some of the ones currently available are deleted.

To modify group-level quotas, the same command is used but with the option -g (edquota -g students).

By default it is to modify quotas for that user in all file systems that have quotas control active (quotaon). If quota control is desired for a specific filesystem then the -f option is added:

```
# edquota -u alumno1 -f /home
```
(only the fee applies in the indicated file system)

STEP 1: Set the parameters in the /etc/fstab file, create quotas for users, groups or both.

Adding the quota to the / etc / fstab file is added to the options field, as a parameter at the user or group quota level.

a)   Edit the file and add the following changes in the / etc / fstab line.

```
# nano /etc/fstab
/dev/sda3  /home    ext3   noatime,usrquota,grpquota  1       2
```

STEP 2: Verify with the commando quotacheck.

Create, verify or repair quota control in the systems that support it.

> quotacheck –augmv

It must be executed periodically.

> quotacheck –ugmv /home

a)   The system is ready to manipulate user quotas, we can verify this because the file "aquota.user" and "aquota.group" must exist at the root of the file system supported with quotas..

> cd /home
>
> ls –l

NOTE: quota.user and quota.group are binary files that should not be edited or modified.

If there are more file systems with quotas at the root of each one, there would be the files, or only one depending on what was requested, users, groups or both.

In systems with kernel 2.2 or earlier, version 1 of quotas was used and its control files were named "quota.user" and "quota.

From here the system is ready to support quotas.

STEP 3: Activate and deactivate disk quotas.

If they are still not activated, it is necessary to activate quota support, for which we invoke the command.

> **quotaon**

a)   Activate the user and group quota in the /home directory.

> # quotaon -ugv /home
>
> /dev/sda3 [/home]: group quotas turned on
>
> /dev/sda3 [/home]: user quotas turned on

b)   Disabling the disk quota.

> **quotaoff**
>
> **# quotaoff -v /home**
>
> **/dev/sda3 [/home]: group quotas turned off**
>
> **/dev/sda3 [/home]: user quotas turned off**

STEP 4: Verify the use of quotas.

The administrator 'root' can see the use of quotas of any user, either individually or through a global summary.

> quota

a) Check a user's fees

```
# quota -u  alumno1
Disk quotas for user alumno1 (uid 1002):
      Filesystem  blocks    quota    limit    grace    files    quota    limit    grace
      /dev/sda3      56       70      100                14        0        0
```

b) See users who have assigned the high quotas, it is a bit difficult to calculate in terms of megabytes or gigas the space used and the quota limits:

```
# quota -u Juan
Disk quotas for user Juan (uid 1001):
   Filesystem blocks  quota limit grace  files quota limit grace
   /dev/sda3 42578888   0 50000000        34895   0   0
```

c) Using the -s option improves the report.

```
# quota -s -u Juan
Disk quotas for user Juan (uid 1001):
   Filesystem blocks  quota limit grace  files quota limit grace
   /dev/sda3 41582M    0 48829M          34905   0   0
```

d) See the quotas as user, without parameter.

```
quota
```

e) See global report of the quotas of all users or groups, such as "root".

```
# repquota /home
*** Report for user quotas on device /dev/sda3
Block grace time: 7days; Inode grace time: 7days
                       Block limits            File limits
   User            used    soft    hard  grace  used  soft  hard  grace
   ----------------------------------------------------------------
   root        --  184280     0       0          11     0     0
   Juan        -- 42579852    0 50000000       34902    0     0
   alumno1     --      56    70     100          14     0     0
   alumno2     --      52     0       0          13     0     0
   alumno3     --      28     0       0           7     0     0
   alumno4     --      28     0       0           7     0     0
```

f) See sizes.

```
repquota  -s
```

g) View a report for all file systems that support quotas.

```
repquota    -a
```

h) If you want to add a report by groups, by default it will be displayed by users.

```
repquota  -g
```

> NOTE: Thanks times per user must be less than global. And that it starts running once the soft limit has been reached.

STEP 5: Establish a time of grace.

At a global level, a grace period for everyone, use the -t option of the edquota command, as in the following example, remember that you must be "root"

> NOTE: if you enter to edit the user's grace time again (edquota -u user -T) it will be reflected in seconds the time it has left, and can increase it again if you are "root".
> If left to zero, the global is used

a) Set the 7 days is the default period, if you change it to say 12 hours, it would be "12hours". The grace time may be different for the soft limit by blocks or by inodes.

```
# edquota -t
Grace period before enforcing soft limits for users:
Time units may be: days, hours, minutes, or seconds
  Filesystem              Block grace period      Inode grace period
  /dev/sda3                    7days                    7days
```

b) Assign the default time to a specific user.

```
# edquota -u alumno1 -T
Times to enforce softlimit for user alumno1 (uid 1002):
Time units may be: days, hours, minutes, or seconds
  Filesystem                       block grace            inode grace
  /dev/sda3                          unset                  unset
```

STEP 6: Set quotas globally to all users.

On Linux operating systems if you have few users, establish user fees per user no problem. But if we speak for example of a university | institute | public administration, where there could be thousands of accounts then it is a problem to establish accounts individually. There is not really an "official" way to establish quotas massively, however, there is no problem, we will use a small script that will allow you to do it.

a) Set the quota you want globally in a single user.

```
# edquota -u user1
Disk quotas for user user1 (uid 2002):
  Filesystem                blocks      soft      hard    inodes    soft     hard
  /dev/sda3                     68       300       400        17       0        0
:wq
```

b) Display a quota summary.

```
Repquota
[root@baldo ~]# repquota /home
*** Report for user quotas on device /dev/sda3
Block grace time: 7days; Inode grace time: 7days
                  Block limits            File limits
```

User		used	soft	hard	grace	used	soft	hard	grace
user1	--	68	300	400		17	0	0	
user2	--	352	0	0		13	0	0	
user3	--	28	0	0		7	0	0	
user4	--	28	0	0		7	0	0	

c) Then we will use the option -p (protptype) to make duplicates from the already established.

```
# edquota -p user1 user2
```

d) Copy the quota limits information from "user1" to "user2", there is no limit on how many users you can place as arguments.

```
# edquota -p user1 user2 user3 user4
```

> NOTE: Only the user "user1" has quotas, the "grace" columns will have values once the soft or soft limit is reached.

e) For a few users but useless if we need to duplicate it in hundreds of users, so let's make a compound command that extracts the names of the users, you can use for example gawk or awk.

```
# gawk -F: '$3 > 999 {print $1}' /etc/passwd
user1
user2
user3
user4
```

> NOTE: The separator ":" of fields (-F), and we indicate as an action that in field 3 ($ 3) we look for all the UIDs greater than 1009 and that print them ({print $ 1}).

f) Preceding gwak for the command along with edquota -p

```
# edquota -p user1 `gawk -F: '$3 > 999 {print $1}'
/etc/passwd`
```

g) Visualize the days of grace and the exceeding of the number of soft and hard blocks exceeded.

```
# repquota /home
*** Report for user quotas on device /dev/sda3
Block grace time: 7days; Inode grace time: 7days
```

		Block limits				File limits			
User		used	soft	hard	grace	used	soft	hard	grace
user1	--	68	300	400		17	0	0	
user2	--	352	300	400	7days	13	0	0	
user3	--	28	300	400		7	0	0	
user4	--	28	300	400		7	0	0	

All users have the same quotas as the "user1" that was the prototype for the others and second it is observed that the "user" that has 352 blocks used when passing the soft limit entered the grace period automatically that the global is 7 days. At the moment when the limit exceeded 300, the grace period began. Now you can only create more files for 7 days or when you get to 400, the first thing that happens, of course, assuming you do not erase first files to recover space.

> NOTE: The "user2" has not reached the "hard" limit nor has the grace time expired, the system allows you to create the file but you are notified with a warning.

STEP 7: Display posted quota notices.

```
warnquota
```

a) Display the error message when listing the file system of a directory. When a user reaches the soft or soft limit when creating or modifying a document, a similar message appears, which is shown here.

```
baldo> ls -l > directorio.txt
sda3: warning, user block quota exceeded.
```

b) Review the file systems with active quotas (quotaon). Check all users, looking for who has exceeded the limit of soft both blocks and inodes, the user who exceeded is sent a notification email.

```
warnquota
```

c) Schedule the task as a job to be done every 12 hours.

```
# nano /etc/crontab
...
0 0,12 * * * root /usr/sbin/warnquota
...
```

> NOTE warnquota comes with the messages in English by default, the file "is /etc/warnquota.conf", is very intuitive and easy to change, personalize it with messages to Spanish to make it easier to understand than your users who have exceeded their dues

PRACTICE 3: Partitions file systems with MBR and GPT.

DESCRIPTION:

The Linux file systems, there is only one root directory and the mount point of the file system, in text environment are made in /mnt and in graphic environment in /media.

By default, the file systems that are mounted in the boot sequence are /etc/fstab.

When the file system is mounted, it depends on how it is done:

> / --> root directory in a partition.
>
> swap -> file system exchange in another partition. (1-2 "4").
>
> 16 Gb of RAM -> 2 SWAP (4 Gb)

Depends on how they want to install.

> / --> partition
>
> /usr --> p
>
> /bin --> p
>
> /mnt

> NOTE: There are Linux versions, which are the development pattern, or origin or design matrix of the rest of the versions are in principle three: Debian, Slackware, RedHAT.

STEP 1: Open a Windows file system from Linux.

a) Start from an ISO.

> Xubuntu.ISO

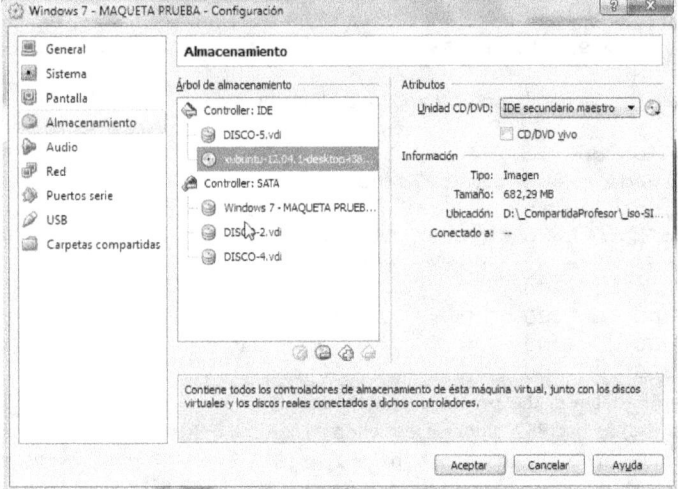

To accept

> Start MV
>
> > Start graphic environment.

STEP 2: Go to text environment.

Open console 1 (tty1).

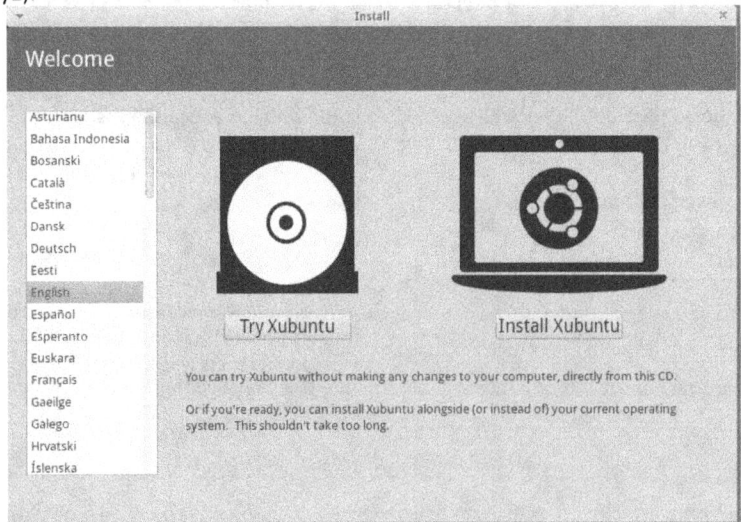

> CTRL+ALT+F1 ... F6
>
> CTRL+ALT+F7 ---> Graphic environment

STEP 3: Work in superuser mode.

> su
>
> $ > user
>
> # root

a) Work on the Line orders.
        sudo command
        sudo su
b) In superuser mode, it is indexed at the prompt with the symbol #.
        #
b.1) The root does not have passwd.
        sudo passwd   root
        > passwd usuario actual
        > passwd root (2 times)
                su   (user root)
                  : passwd
b.2) Skip login.
        $ sudo su -l

STEP 4: View the mounted file systems.
        df
        cat /etc/fstab
  /etc/fstab: It is used to define how partitions, different block devices or remote file systems must be mounted and integrated into the system.
  Each file system is described on a separate line. These definitions will be converted with systemd into dynamically mounted units at startup, and when the system administrator's configuration is reloaded.
  The file is read by the mount command, which is enough to find any of the directories or devices indicated in the file to complete the value of the next parameter. By doing so, the mounting options that are listed in fstab will also apply.

STEP 5: Display the connected disks.
        fdisk   -l
Text console management.
        MAY +   RePag

```
Disk /dev/sda: 37.6 GB, 37580963840 bytes
255 heads, 63 sectors/track, 4568 cylinders, total 73400320 sectors
Units = sectors of 1 * 512 = 512 bytes
Sector size (logical/physical): 512 bytes / 512 bytes
I/O size (minimum/optimal): 512 bytes / 512 bytes
Disk identifier: 0xae4c7688

   Device Boot      Start         End      Blocks   Id  System
/dev/sda1   *        2048      206847      102400    7  HPFS/NTFS/exFAT
/dev/sda2          206848    73398271    36595712    7  HPFS/NTFS/exFAT
```

Windows, when installing the system:
Create a partition for the boot manager.
For Windows it is a system partition.
  • Windows xp 8 Mb.
  • Windows 7 100 Mb.
  • Windows 8 300-340 Mb.
The partition where the OS is. -> The MAIN partition.
a) Access the partition.
        fdisk   /dev/sda
b) Access to the fdisk application.
        m   help.

```
root@xubuntu:~# fdisk /dev/sda

Command (m for help): m
Command action
   a   toggle a bootable flag
   b   edit bsd disklabel
   c   toggle the dos compatibility flag
   d   delete a partition
   l   list known partition types
   m   print this menu
   n   add a new partition
   o   create a new empty DOS partition table
   p   print the partition table
   q   quit without saving changes
   s   create a new empty Sun disklabel
   t   change a partition's system id
   u   change display/entry units
   v   verify the partition table
   w   write table to disk and exit
   x   extra functionality (experts only)
```

b.1) Display partitions Press the letter p + [ENTER]

```
Command (m for help): p

Disk /dev/sda: 37.6 GB, 37580963840 bytes
255 heads, 63 sectors/track, 4568 cylinders, total 73400320 sectors
Units = sectors of 1 * 512 = 512 bytes
Sector size (logical/physical): 512 bytes / 512 bytes
I/O size (minimum/optimal): 512 bytes / 512 bytes
Disk identifier: 0xae4c7688

   Device Boot      Start         End      Blocks   Id  System
/dev/sda1   *        2048      206847      102400    7  HPFS/NTFS/exFAT
/dev/sda2          206848    73398271    36595712    7  HPFS/NTFS/exFAT
```

b.2) Exit or quit fdisk.

    q

    fdisk -l

a.)   Create a new partition (1..4) MBR.

    fdisk /dev/sdd

      Create an extended partition (one per disk), from the number (5 ....), the logical partitions are assigned.

          MBR --> Primary or extended.

          EMBR --> logical   (5...)

              n   new

                  p   primary

                  1

                  First sector [ENTER]

                      +5G

        +5120M

      n

          p primary

             2

      First sector   [ENTER]

2G

```
Disk /dev/sdd: 26.8 GB, 26843545600 bytes
255 heads, 63 sectors/track, 3263 cylinders, total 52428800 sectors
Units = sectors of 1 * 512 = 512 bytes
Sector size (logical/physical): 512 bytes / 512 bytes
I/O size (minimum/optimal): 512 bytes / 512 bytes
Disk identifier: 0xe4548064

   Device Boot      Start         End      Blocks   Id  System
/dev/sdd1           2048    10487807     5242880   83  Linux
/dev/sdd2       10487808    14682111     2097152   83  Linux
```

d.)   Change the type of file system.

    t

    Number: 2   partition2

        l --> Display the file system table.

e.)   Assign partition 82 and display the file system table.

    82

        Display p

f.)   Set active partition.

    a --> Set / delete the active partition.

        Number partition: 1

        p --> Display table partition

> Change the file system.

```
Disk /dev/sdd: 26.8 GB, 26843545600 bytes
255 heads, 63 sectors/track, 3263 cylinders, total 52428800 sectors
Units = sectors of 1 * 512 = 512 bytes
Sector size (logical/physical): 512 bytes / 512 bytes
I/O size (minimum/optimal): 512 bytes / 512 bytes
Disk identifier: 0xe4548064

   Device Boot      Start         End      Blocks   Id  System
/dev/sdd1   *        2048    10487807     5242880   83  Linux
/dev/sdd2       10487808    14682111     2097152   82  Linux swap / Solaris
```

g.)   Save table and exit fdisk.

    w

h.)   Extended partition.

    fdisk   /dev/sdd

    Logical partition.

        n

            e

    Assign partition size

        The first value is allowed to be assigned by default [ENTER]

        The last value is the completion of the partition is usually assigned a size M, G, T: +15G

i.) Delete a partition.

Select a partition and press the d: delete key.

d

j.) Check or verify the table.

v ---> verify

```
Command (m for help): v
Remaining 37748734 unallocated 512-byte sectors
```

STEP 6: Enter expert mode.

We access the sdd partition and the expert mode by pressing x, change the options and consult the help with m. We view the layout of the partition table with MBR or EMBR system, we can see the 4 inputs, the Cylinders, Heads and Sectors.

fdisk /dev/sdd

x -->   There is a menu

m --> List the expert mode command options. (HELP)

```
Expert command (m for help): m
Command action
   b   move beginning of data in a partition
   c   change number of cylinders
   d   print the raw data in the partition table
   e   list extended partitions
   f   fix partition order
   g   create an IRIX (SGI) partition table
   h   change number of heads
   i   change the disk identifier
   m   print this menu
   p   print the partition table
   q   quit without saving changes
   r   return to main menu
   s   change number of sectors/track
   v   verify the partition table
   w   write table to disk and exit

Expert command (m for help): _
```

p --> Visualize the partition table.

```
Expert command (m for help): p

Disk /dev/sdd: 255 heads, 63 sectors, 3263 cylinders

Nr AF  Hd Sec  Cyl  Hd Sec  Cyl     Start      Size ID
 1 80  32  33    0 213   9  652      2048  10485760 83
 2 00 213  10  652 234  25  913  10487808   4194304 82
 3 00 234  26  913  11  19  824  14682112  31457280 05
 4 00   0   0    0   0   0    0         0         0 00
 5 00   0   0    0   0   0    0         0         0 00

Expert command (m for help): _
```

h --> Head.

200

```
Expert command (m for help): h
Number of heads (1-256, default 255): 200

Expert command (m for help): p

Disk /dev/sdd: 200 heads, 63 sectors, 3263 cylinders

Nr AF  Hd Sec  Cyl  Hd Sec  Cyl     Start      Size ID
 1 80  32  33    0 213   9  652      2048  10485760 83
 2 00 213  10  652 234  25  913  10487808   4194304 82
 3 00 234  26  913  11  19  824  14682112  31457280 05
 4 00   0   0    0   0   0    0         0         0 00
 5 00   0   0    0   0   0    0         0         0 00
```

Check different partitions file systems

fsck
fsck.cramfs
fsck.ext2 -> e2fsck
fsck ext3 -> e2fsck
fsck ext4 -> e2fsck
fsck ext4dev -> e2fsck
fsck.fat
fsck.minix
fsck.msdos -> fsck.fat
fsck.nfs
fsck.vfat -> fsck.fat
fsfreeze
fstab-decode
fstrim
fstrim-all

STEP 7: Check file systems of a Linux and Windows.

The orders are in the / sbin directory, the first thing we do is visualize the orders.

ls -l   fs*

fsck

Pre-check the file systems that exist for each disk.

fdisk -l

a)   Help.

fsck --help

b)   Check Linux file systems ext.

fsck -p -f /dev/sda1

c)   Check a Windows file system.

fsck.msdos -a   /dev/sdb1

fsck.ntfs   -a   /dev/sdb2

fsck.vfat   -a   /dev/sdb3

STEP 8: Access GPT partitions.

The commands that are used from the command line are: gdisk and cgdisk, available for 64 bits, the practices are made with slackware that incorporates the two orders.

gdisk -> basic in management, very broad.

cgdisk -> basic menu.

a) Access to visualize the disk partitions /dev/sdd

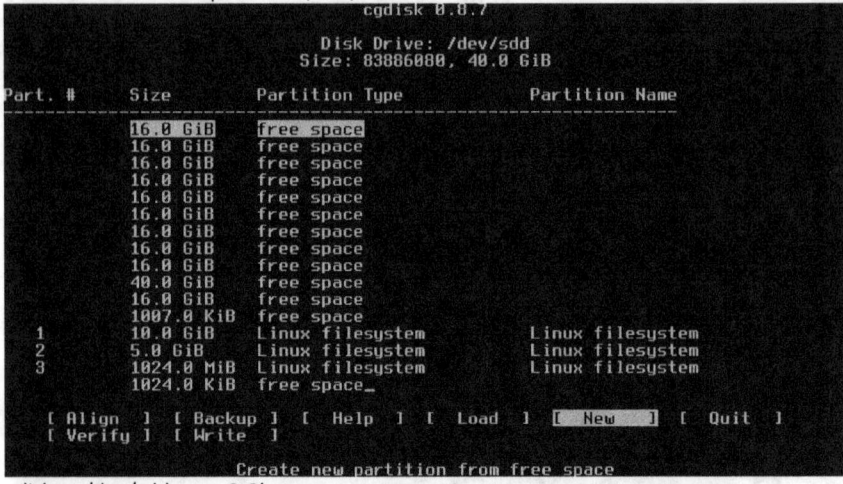

gdisk   /dev/sdd   --> 8 Gbytes

q --> quit

gdisk /dev/sdd

? --> help

b) Visualize the partitions.

```
p
Disk /dev/sdd: 83886080 sectors, 40.0 GiB
Logical sector size: 512 bytes
Disk identifier (GUID): 4C28E8BD-5677-4E78-9D6A-E1EE795CCBA2
Partition table holds up to 128 entries
First usable sector is 34, last usable sector is 83886046
Partitions will be aligned on 2048-sector boundaries
Total free space is 83886013 sectors (40.0 GiB)

Number  Start (sector)   End (sector)  Size        Code  Name
```

c) Delete partitions.

The normal thing is to erase the logical partitions, when they have been erased all the extended partition is deleted, and then the primary partitions are erased.

Delete the extended -> there is extended in MBR

GPT -> There are no extended partitions, all are logical.

You have to erase them all, one by one

STEP 9: Create GPT partitions.

gdisk /dev/sdd

a) Once we have accessed Show help.

Command (? for help): **?**

b) Create a new GPT partition.

n --> new partition

1..128 --> 1

TYPE (P/LOGICAL) --> IT DOES NOT EXIST, THEY ARE ALL EQUALLY

[FIRST SECTOR] <ENTER>

[LAST SECTOR] +4G

All new types of file systems are shown as 16 bits, 2 bytes are used. From 0000 to FFFF.

```
Command (? for help): n
Partition number (1-128, default 1):
First sector (34-83886046, default = 2048) or {+-}size{KMGTP}:
Last sector (2048-83886046, default = 83886046) or {+-}size{KMGTP}: _
Hex code or GUID (L to show codes, Enter = 8300): L
0700 Microsoft basic data  0c01 Microsoft reserved   2700 Windows RE
4200 Windows LDM data       4201 Windows LDM metadata 7501 IBM GPFS
7f00 ChromeOS kernel        7f01 ChromeOS root        7f02 ChromeOS reserved
8200 Linux swap             8300 Linux filesystem     8301 Linux reserved
8400 Intel Rapid Start      8e00 Linux LVM            a500 FreeBSD disklabel
a501 FreeBSD boot           a502 FreeBSD swap         a503 FreeBSD UFS
a504 FreeBSD ZFS            a505 FreeBSD Vinum/RAID    a580 Midnight BSD data
a581 Midnight BSD boot      a582 Midnight BSD swap     a583 Midnight BSD UFS
a584 Midnight BSD ZFS       a585 Midnight BSD Vinum    a800 Apple UFS
a901 NetBSD swap            a902 NetBSD FFS            a903 NetBSD LFS
a904 NetBSD concatenated    a905 NetBSD encrypted      a906 NetBSD RAID
ab00 Apple boot             af00 Apple HFS/HFS+        af01 Apple RAID
af02 Apple RAID offline     af03 Apple label           af04 AppleTV recovery
af05 Apple Core Storage     be00 Solaris boot          bf00 Solaris root
bf01 Solaris /usr & Mac Z   bf02 Solaris swap          bf03 Solaris backup
bf04 Solaris /var           bf05 Solaris /home         bf06 Solaris alternate se
bf07 Solaris Reserved 1     bf08 Solaris Reserved 2    bf09 Solaris Reserved 3
bf0a Solaris Reserved 4     bf0b Solaris Reserved 5    c001 HP-UX data
c002 HP-UX service          ea00 Freedesktop $BOOT     eb00 Haiku BFS
ed00 Sony system partitio   ef00 EFI System            ef01 MBR partition scheme
ef02 BIOS boot partition    fb00 VMWare VMFS           fb01 VMWare reserved
fc00 VMWare kcore crash p   fd00 Linux RAID
Hex code or GUID (L to show codes, Enter = 8300): _
c002 HP-UX service          ea00 Freedesktop $BOOT     eb00 Haiku BFS
ed00 Sony system partitio   ef00 EFI System            ef01 MBR partition scheme
ef02 BIOS boot partition    fb00 VMWare VMFS           fb01 VMWare reserved
fc00 VMWare kcore crash p   fd00 Linux RAID
Hex code or GUID (L to show codes, Enter = 8300): l
0700 Microsoft basic data  0c01 Microsoft reserved   2700 Windows RE
4200 Windows LDM data       4201 Windows LDM metadata 7501 IBM GPFS
7f00 ChromeOS kernel        7f01 ChromeOS root        7f02 ChromeOS reserved
8200 Linux swap             8300 Linux filesystem     8301 Linux reserved
8400 Intel Rapid Start      8e00 Linux LVM            a500 FreeBSD disklabel
a501 FreeBSD boot           a502 FreeBSD swap         a503 FreeBSD UFS
a504 FreeBSD ZFS            a505 FreeBSD Vinum/RAID    a580 Midnight BSD data
a581 Midnight BSD boot      a582 Midnight BSD swap     a583 Midnight BSD UFS
a584 Midnight BSD ZFS       a585 Midnight BSD Vinum    a800 Apple UFS
a901 NetBSD swap            a902 NetBSD FFS            a903 NetBSD LFS
a904 NetBSD concatenated    a905 NetBSD encrypted      a906 NetBSD RAID
ab00 Apple boot             af00 Apple HFS/HFS+        af01 Apple RAID
af02 Apple RAID offline     af03 Apple label           af04 AppleTV recovery
af05 Apple Core Storage     be00 Solaris boot          bf00 Solaris root
bf01 Solaris /usr & Mac Z   bf02 Solaris swap          bf03 Solaris backup
bf04 Solaris /var           bf05 Solaris /home         bf06 Solaris alternate se
bf07 Solaris Reserved 1     bf08 Solaris Reserved 2    bf09 Solaris Reserved 3
bf0a Solaris Reserved 4     bf0b Solaris Reserved 5    c001 HP-UX data
c002 HP-UX service          ea00 Freedesktop $BOOT     eb00 Haiku BFS
ed00 Sony system partitio   ef00 EFI System            ef01 MBR partition scheme
ef02 BIOS boot partition    fb00 VMWare VMFS           fb01 VMWare reserved
fc00 VMWare kcore crash p   fd00 Linux RAID
Hex code or GUID (L to show codes, Enter = 8300): _
```

b) Create a new Swap partition.

    n

    2

    [ENTER]

    +2G

    8200

c) Visualize the partitions.

    p

d) Create a partition with all the available space.

    n

    3

    [ENTER]

    [ENTER]

    8300

e) Set the active partition.

    ?

    x --> Expert Mode.

```
Expert command (? for help): ?
a    set attributes
c    change partition GUID
d    display the sector alignment value
e    relocate backup data structures to the end of the disk
g    change disk GUID
h    recompute CHS values in protective/hybrid MBR
i    show detailed information on a partition
l    set the sector alignment value
m    return to main menu
n    create a new protective MBR
o    print protective MBR data
p    print the partition table
q    quit without saving changes
r    recovery and transformation options (experts only)
s    resize partition table
t    transpose two partition table entries
u    Replicate partition table on new device
v    verify disk
w    write table to disk and exit
z    zap (destroy) GPT data structures and exit
?    print this menu
```

NOTE: SWAP once formatted is activated, but it is not a START partition with what should not be set as ACTIVE

a --> set attributes.

      3 [ENTER]

```
Expert command (? for help): a
Partition number (1-3): 3
Known attributes are:
0: system partition
1: hide from EFI
2: legacy BIOS bootable
60: read-only
62: hidden
63: do not automount

Attribute value is 0000000000000000. Set fields are:
   No fields set
```

0 --> SYSTEM PARTITION

[ENTER]

w --> save.

f) List the partitions of a GPT unit.

    gdisk /dev/sdd

```
Disk /dev/sdd: 42.9 GB, 42949672960 bytes
256 heads, 63 sectors/track, 5201 cylinders, total 83886080 sectors
Units = sectors of 1 * 512 = 512 bytes
Sector size (logical/physical): 512 bytes / 512 bytes
I/O size (minimum/optimal): 512 bytes / 512 bytes
Disk identifier: 0x00000000

   Device Boot      Start         End      Blocks   Id  System
/dev/sdd1               1    83886079    41943039+  ee  GPT
```

WORK UNIT II: Directories on Linux

- *PRACTICE 4: User and group key files.*

- *PRACTICE 5: Manage the different Shell.*

- *PRACTICE 6: Manage directories in Linux.*

- *PRACTICE 7: Handle the most common apt-get options.*

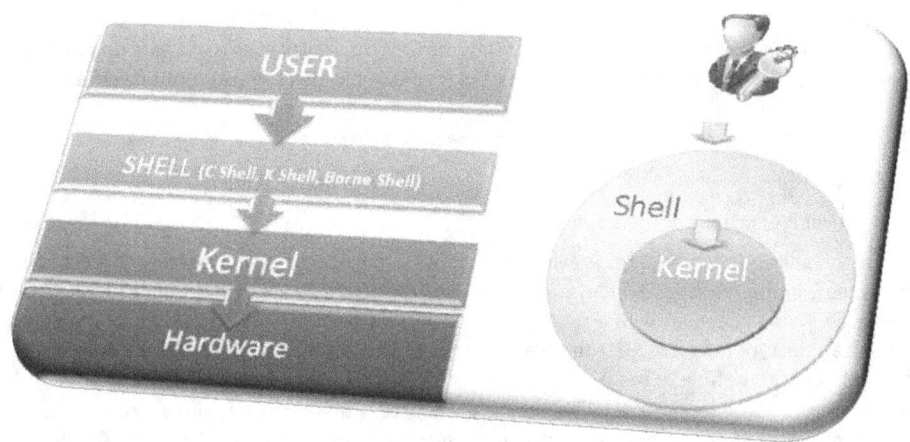

Contents
- Basic commands in Linux.
- Directories in Linux.
- The File System in Linux.
- Help in Linux.
- Operations on directories and folders.
- Attributes of directories or folders.

Command

/etc/passwd
/etc/group
/etc/shadow
/etc/gshadow
sudo, info, infotext,
cat, init, halt,
poweroff, reboot,
reset, clear, pwd,
cd, man, fdisk,
echo, tree, mkdir,
rm, touch, mv,
apt-get, shutdown,
chown, chmod,
mknod,
dpkg-reconfigure,
dpkg

PRACTICE 4: User and group key files.

DESCRIPTION:

The files that form the configuration of users and groups are in the / etc directory, and are:

Administration and user control files

ADMINISTRATION FILES AND CONTROL OF USERS	FUNCTIONALITY
.bash_logout	It is executed when the user leaves the session.
.bash_profile	It is executed when the user starts the session.
.bashrc	Run when the user starts the session
/etc/group	Users and their groups.
/etc/gshadow	Encrypted passwords of the groups.
/etc/login.defs	Variables that control the aspects of user creation.
/etc/passwd	System users.
/etc/shadow	Encrypted passwords and date control of system users.

/etc/passwd

The /etc/passwd file contains most of the user account information. This information is available to all users on most systems with just using cat /etc/passwd, but only the root user can modify it. This file exists in FreeBSD, but there is also /etc/master.passwd with the same information.

login ID : x : UID number : número de grupo : Comentarios : Directorio de trabajo : Shell de usuario

FIELD	DESCRIPTION
login ID	ID is the name with which the account is accessed.
x	Represents the encrypted password. Previously the encrypted password appeared in this section, but for security reasons it is now in the /etc/shadow file. Some Unix versions may still include it, but in general it is something that is no longer used. Remember that with a simple cat /etc/passwd any user has access to the encrypted code, and with brute force can decrypt it. In FreeBSD the file /etc/master.passwd does contain the encrypted passwords, but you need root privileges to be able to see it.
UID number Group number	The user identification number. For convenience, users access their account with a name chosen by them; but for Unix users are represented by a number that in most systems goes from 0 to 65535, with 0 - 99 reserved for system files. This identification number can be duplicated by the administrator, although there may be confusion and it is not recommended. The root user has reserved the number 0. As what really matters for Unix is not the name without or the ID, then any user with a number 0 has root privileges.
Número de grupo	Group number by default. Represents the group to which the user is assigned at the beginning. This number is not unique, and many users can share it without problems.
Comments ..	Comments and additional data of the account. It includes general information that is requested when creating the user. This field can be blank. It is also not convenient to include sensitive information, because everyone will be able to see it. This field is known as GCOS (from the General Electric Comprehensive Operating System).
Working directory	Directory in which the session is initiated by default. Generally this field contains something like / home / username, indicating that the student account is mounted in the home directory. It is not necessary to be in that directory, but the account must be avoided in: / temp.
The user shell	User login shell. It needs to be one of those contained in the /etc/shell file. Each of these fields is separated by ':' (two dots) If any of these fields is empty, they will appear (two dots twice).

Ej.:

```
# cat /etc/password
root:x:0:0:root:/root:/bin/bash
bin:x:1:1:bin:/bin:/bin/sh
daemon:x:2:2:daemon:/sbin:/bin/sh
```

/etc/shadow

user: password: last: can: must: notice: expires: deactivates: reserved

FIELD	DESCRIPTION
user	The name of the user..
password	The encrypted key.
last	Days elapsed since the last change of password from day 1/1/70.
can	Days elapsed before the key can be modified.
must	Days elapsed before the key has to be modified.
notice	Days of warning the user before the key expires.
expires	Days the account is deactivated after the key expires
desactivate	Days of duration of the account from 1/1/70.
reserved	No comment.

Ej.: # cat /etc/shadow
victoria : gEvm3sslnGRlr : 10639 : 0 : 99999 : 7 : -1 : -1 : 134529868
alumno:6h5osz0oA$BLZlWenCbtcK9tP060Med5XTgSZ53ziCzQvAmTb2DAbRmlrwM4FnQ/NH80jBuZm8jdo.d3tA1L4vaDTSJ6pbf
1:16210:0:99999:7:::
admin:6vlsmtqCx$1V9/lDQ7NoF3EBzwJ8aFrJbjeqD.wiEVNl0xrQ/VrPsxvL28SJCHrAv3ipqeGBnnWOP99bQV1Dg3OqeMrphGw1:
16237:0:99999:7:::

/etc/group

grup:password:gid:users

It contains the names of the groups and a list of the users belonging to each group.

FIELD	DESCRIPTION
grup	The name of the group (it is recommended that it not have more than 8 characters): samba share
password	The encrypted password or an x that indicates the existence of a gshadow file: x
gid	The GID number of the group (group identification number): 124.
users	List of members of the group, separated by commas (without spaces): student.

By default the one belonging to the group that is defined in / etc / passwd will prevail in case of disagreeing with this file.
Ej.:   # cat /etc/group
sambashare:x:124:alumno

/etc/gshadow

As with the /etc/shadow file of encrypted passwords for users, you can also use a /etc/gshadow file of encrypted passwords for groups.

name: password: uid: gid: optional description folder: shell

FIELD	DESCRIPTION
name	Numbers are not allowed at the beginning of a username.
password	An "x" indicates that the password is stored in / etc / shadow, in the case of being an "!" It is that the user is blocked. If you have "!!" it is that you do not have it.
uid	Each user carries a non-identifier (uid) between 0 (root) and 65535. Some are reserved for the root user (always zero), and for users of various services in the system. Red Hat and derivatives between 1 and 499. Debian and derivatives between 1 and 999.
gid	Group id, each user has a main group id, but can belong to more groups.
folder	It will use it as the user's home folder, when logging in with it, it will be the one that loads by default.
shell	Users of services and users with limited permissions should not have a shell, that is, log in to the console, they are usually left with /usr/bin/nologin or /bin/false

Ej.:   # cat /etc/gshadow
lpadmin:!::alumno
scanner:!::saned
alumno:!::
sambashare:!::alumno

It is usually used to allow access to the group, to a user who is not a member of the group. That user would then have the same privileges as the members of his new group.

/usr/sbin/pwconv   To convert to shadow format.
/usr/sbin/pwunconv   To convert back to the traditional format.

pwconv y pwunconv

The default behavior of all modern GNU/Linux distributions (distros) is to enable extended protection of the /etc/shadow file, which (insists) effectively hides the encrypted 'hash' of the /etc/passwd password.

But if for some bizarre and strange compatibility situation it were required to have the passwords encrypted in the same file of /etc/passwd, the pwunconv command would be used:

#> more /etc/passwd
root:x:0:0:root:/root:/bin/bash
sergio:x:1001:1000:Sergio González:/home/sergio:/bin/bash
...

(The 'x' in field 2 indicates that use is made of / etc / shadow).

#> more /etc/shadow
root:ghy675gjuXCc12r5gt78uuu6R:10568:0:99999:7:7:-1::
sergio:rfgf886DG778sDFFDRRu78asd:10568:0:-1:9:-1:-1::
#> pwunconv
#> more /etc/passwd
root:ghy675gjuXCc12r5gt78uuu6R:0:0:root:/root:/bin/bash
sergio:rfgf886DG778sDFFDRRu78asd:1001:1000:Sergio González:/home/sergio:/bin/bash
...
#> more /etc/shadow
/etc/shadow: No such file or directory

(When running pwunconv, the shadow file is deleted and the encrypted passwords 'passed' to passwd). At any time it is possible to reactivate shadow protection:

```
#> pwconv
#> ls -l /etc/passwd /etc/shadow
•     rw-r—r-- 1 root root 1106 2007-07-08 01:07 /etc/passwd
•     r-------- 1 root root  699 2009-07-08 01:07 /etc/shadow
```

The shadow file is re-created, also note the very restrictive permissions (400) that this file has, making it extremely difficult (I do not like to use impossible, since in computer it seems that the impossible 'almost' do not exist) that any user that is not root, read it.

/etc/login.defs

In the configuration file /etc/login.defs are defined the variables that control the aspects of user creation and shadow fields used by default. Some of the aspects that control these variables are:
- Maximum number of days that a password is valid PASS_MAX_DAYS.
- The minimum number of characters in the password PASS_MIN_LEN.
- Minimum value for normal users when using useradd UID_MIN.
- The default Umask value UMASK.
- If the command useradd must create the home directory by default CREATE_HOME.

Just read this file to know the rest of the variables that are self-descriptive and adjust them to taste. Remember that they will be used mainly when creating or modifying users with the useradd and usermod commands that will be explained shortly.

Content of the hidden files found in the user part HOME.
Example:
List of summary commands:

COMMAND	DESCRIPTION
chage	Change the information on the expiration of the password.
chfn	Change the information in the "comment" field of a user.
chsn	Change the information in the "shell" field of a user.
groupadd	Add groups to the system.
groupdel	Delete a group that exists.
groupmod	Modify the parameters of a group existing in the system.
groups	It says in which groups we are.
id	Shows ID and groups.
login	It allows to change the user.
mesg	[and / n] allow or not to write messages to you.
newgrp	It allows to change to another group (we need to know the password).
sg	Allows you to execute commands from another group.
su	It allows to change to superuser (root).
talk	Interactive bidirectional communication with another user that is connected to the system.
useradd	Add users to the system.
userdel	Delete users.
usermod	Modify the parameters of a user.
w	List the users that are in the system and what they are doing.
wall	Write message to all users.
who	List the users that are in the system.
whoami	It says which user we are.
write	Write a message to another user.

PRACTICE 5: Manage the different Shell.

DESCRIPTION:

Display modes of the shell is the command interpreter (equivalent to cmd) there are different versions of shell that can be divided into four categories: Bourne type, C console type, non-traditional and historical.

Compatible with Bourne Shell.

- **Bourne shell (sh):** Written by Steve Bourne, when he was in Bell Labs. It was distributed for the first time with Version 7 Unix, in 1978, and improved over the years.
- **Almquist shell (ash):** Written as a replacement for the Bourne shell with BSD license; FreeBSD sh, NetBSD (and its derivatives) are based on ash and have been improved according to POSIX for the occasion.
- **Bourne-Again shell (bash):** It was written as part of the GNU project to provide it with a superset of functionality with the Bourne shell.
- **Debian Almquist shell (dash):** Dash is a modern replacement of ash in Debian.
- **Korn shell (ksh):** Written by David Korn, while at Bell Labs.
- **Z shell (zsh):** Considered the most complete: it is the closest thing that exists to encompass a superset of sh, ash, bash, csh, ksh, and tcsh.

Display modes of the shell is the command interpreter (equivalent to cmd) there are different shell versions that can be divided into four categories: Bourne type, C console type, non-traditional and historical.

Compatible with Bourne Shell.

- **Bourne shell (sh):** Written by Steve Bourne, when he was in Bell Labs. It was distributed for the first time with Version 7 Unix, in 1978, and improved over the years.
- **Almquist shell (ash):** Written as a replacement for the Bourne shell with BSD license; FreeBSD sh, NetBSD (and its derivatives) are based on ash and have been improved according to POSIX for the occasion.
- **Bourne-Again shell (bash):** It was written as part of the GNU project to provide it with a superset of functionality with the Bourne shell.
- **Debian Almquist shell (dash):** Dash is a modern replacement of ash in Debian.
- **Korn shell (ksh):** Written by David Korn, while at Bell Labs.
- **Z shell (zsh):** Considered the most complete: it is the closest thing that exists to encompass a superset of sh, ash, bash, csh, ksh, and tcsh.

Compatible with the shell of C.

- **C shell (csh)** written by Bill Joy, while at the University of California, Berkeley. It was distributed for the first time with BSD in 1979.
- **TENEX C shell** (tcsh).

Other or exotic.

- **fish:** a friendly and interactive shell, launched for the first time in 2005.
- **mudsh:** a smart shell in the style of video games that operates as a MUD.
- **zoidberg:** a modular shell written in Perl, configured and fully operating in Perl.
- **rc:** the default shell of Plan 9 from Bell Labs and Version 10 of Unix written by Tom Duff. Ports have been made for Inferno and for Unix-based operating systems.
- **it's shell (s):** an RC-compatible shell written in the mid 90's.
- **scsh:** (Scheme Shell)
- There are two types of consoles or two display modes, at the level of the graphic card: the text mode (tty) and the graphic mode (GUI).

Access to consoles.

- CTRL + ALT + F1 ... F6 -> 6 text consoles can be opened simultaneously.
- CTRL + ALT + F7-> Return to the graphical environment.

By default we enter a console at user level, that is reflected in the prompt, for its completion in $.

    $ sudo passwd   root
Password usuario actual: Practical2017*
    Username: root
    Password : Practical2017*

Builtin

SHELL BUILTIN COMMANDS	DESCRIPTION
Bourne Shell Builtins	Builtin commands inherited from the Bourne Shell.
Bash Builtins	Table of builtins specific to Bash.
Modifying Shell Behavior	Builtins to modify shell attributes and optional behavior.
Special Builtins	Builtin commands classified specially by POSIX.

Builtin commands are contained within the shell itself. When the name of a builtin command is used as the first word of a simple command, the shell executes the command directly, without invoking another program. Builtin commands are necessary to implement functionality impossible or inconvenient to obtain with separate utilities.

This section briefly describes the builtins which Bash inherits from the Bourne Shell, as well as the builtin commands which are unique to or have been extended in Bash.

Several builtin commands are described in other chapters: builtin commands which provide the Bash interface to the job control facilities, the directory stack, the command history, and the programmable completion facilities.

Many of the builtins have been extended by POSIX or Bash.

STEP 1: User access.

Login: smr
Password: Practical2017*
.....$ r

Type of user that is accessed is shown by the command line:

$ --> user.
>   --> user.
--> Super user.

Execute orders in superuser mode.

sudo command

> All commands in Linux/Unix are written in lowercase
> Capital letters are reserved for the environment variables.

STEP 2: LINUX syntax.

command parameters arguments

parameters ==   Options or order modifiers (equivalent to Windows options in CMD).
letter

word --literal   A literal is a string of 2 or more characters, normally they are argument words (linux) == parameter (Windows) / path / files == (unit: \ path \ files).

STEP 3: Aid of the orders.

a)   Online help

command --help

Rewind on the console MAY + [EDIT KEYPAD (RePag)]

b)   Different ways of obtaining aid.

b.1) Help using the order man.

man command

Ej.: man ls

b.2) Help using the info order.

Info command

ej.: info ls

b.3) Help using the infotext order.

infotext orden

ej.: infotext ls

b.4) Help using the textinfo command.

textinfo command

ej.: textinfo ls

b.5) List of basic orders in alphabetical order with a brief description.

help

> **Syntax and examples.**
> To exit the help orders q=quit
> Indicates exit from a help application.

STEP 4: Different SHELL.

The operating system has a kernel (it starts the system) and a shell or command interpreter: sh, bash, csh, tcsh, zcsh, zsh, ...

The interpreter allows creating files ("batch batch") Shell or script:

- They are very powerful.
- They allow programming:

Shell Programming Ex: sh, bash.
Users are identified in the command line: $
The root is identified: #
Programming language C: csh, tcs, zcsh Users are
identified in the command line:> The root is identified: #

- The Shell that a user handles is defined when the user is created.
- It is defined in text environment: useradd, adduser, usermod.

> **UUID:** Universally Unique Identifier.
> It allows the existence of different probable devices.
> A UUID is a 16-byte (128-bit) number.
> The theoretical number of possible UUID is then 3×10^{38}
> In its canonical form, a UUID consists of 32 hexadecimal digits, shown in five groups separated by hyphens, of the form 8-4-4-4-12 for a total of 36 characters (32 digits and 4 dashes).

Where is the definition?

The definition is found in /etc/passwd

cat /etc/passwd
clear

STEP 5: File System.

For Linux everything is treated as a file.

With mount point / (root directory ...), the boot partition is mounted.

The rest of the partitions Where is it mounted?

It is assembled from /, as a general rule it is mounted /mnt

/mnt/floppy
/mnt/cdrom
/mnt/dvd
/mnt/Windows

> CONCLUSION: you have to assemble and disassemble the units in a pre-defined assembly point or create the directory, before assembling.

/mnt/pen
mount: mount a file system.
umount: unmount a file system.
  Unit assembly file / ect / fstab
    cat /ect/fstab
Exchange unit, a partition, whose type is / swap
For each partition, corresponding to a storage unit, a UUID identification key appears: universal unique identifier of units.
A file assembly needs a handling device.
    /dev

STEP 6: Treatment of devices at level ..
- Character.
- Block.

Eg: There are times when the MAKEDEV script does not have information about a device, so you have to create a device of character ttySO, character device with number 4 and number 64. Device.txt is the canonical source of the devices.
    # mknod /dev/ttyS0 c 4 64
    # chown root.dialout /dev/ttyS0
    # chmod 0644 /dev/ttyS0
    # ls -l /dev/ttyS0

STEP 7: Stop the operating system.
It can be stopped by closing the virtual machine, or from the command line using any of these commands.

COMMAND	DESCRIPTION
init 0	It is the first running process after loading the kernel and which in turn generates all other processes. Runlevel 0 indicates that the operating system stops and stops.
halt	It is used to turn off the computer.
shutdown	Shut down or restart the system.
power	Turn off the system.
poweroff	Turn off the system.
reboot	Reboot the system.
reset	Reset the Linux system

FILE	DESCRIPTION
/var/run/utmp	File in which the current execution level will be read from; this file will also be updated with the execution level record being replaced by a shutdown time log.
/var/log/wtmp	A new record for the execution level the shutdown time will be appended to this file.

STEP 8: Example of stop and restart of the system.
1. To stop the system:
    halt   This command is similar to the poweroff, which turns off the system.
2. To turn off the system:
    poweroff   The poweroff command is used to shut down the system.
3. To restart the system:
    reboot   The reboot command is used to restart the system.
4. Restart the operating system:
    reset
5. Another way to restart:
    init 6
6. We can ask you to turn off the system right now:
    sudo shutdown -h now
    O well
    sudo shutdown -h +0
7. Request that you turn off the system in a certain time:
    sudo shutdown -h +m
8. Where m is the number of minutes that must elapse before the system shuts down; for example, if we want it to turn off in 10 minutes, it would be:
    sudo shutdown -h +10
9. In Ubuntu it is possible to omit the argument -h leaving only:
    sudo shutdown +10
10. We can also tell you to turn off at a specific time. (It uses the 24-hour system, that is, from 00 to 23). For example at 5:30 p.m.:

halt	
Syntax:	**halt [-d	-f	-h	-n	-i	-p	-w]**
OPTION	DESCRIPTION
-d	Do not write wtmp record (in the file / var / log / wtmp) The -n flag implies –d
-h	Put all system hard drives in standby mode before the system stops or shuts down.
-n	Do not synchronize before restarting or stopping.
-i	Turn off all network interfaces. When you stop the system, you turn it off as well.
-p	This is by default when the halt is called as a poweroff.
-w	Do not restart or stop, just write the wtmp record (in the / var / log / wtmp file).

sudo shutdown -h 17:30

11. In addition you can add a legend to the shutdown command:
   sudo shutdown -h 18:45 "The equipment will be shut down for maintenance"

12. To restart the system:
   sudo reboot
   O well:
   sudo shutdown -r now
   sudo shutdown -r +0

13. To restart the system in a certain time:
   sudo shutdown -r +5

14. To restart the system at a specific time:
   sudo shutdown -r 23:30

sudo	
Syntax:	**sudo [option] [USER]**
OPTION	DESCRIPTION
-b	As --backup but do not accept any argument.
-f	Never ask before overwriting.
-i	Ask for confirmation before overwriting.
-S	Replace the usual backup suffix.
-T	Treat DESTINATION as a normal file.
-u	It only moves when the ORIGIN file is more modern than the destination file.

PRACTICE 6: Handle common commands.
DESCRIPTION:

For Linux all are files, there are no directories or units, no devices, everything is treated as a file system.

However, we must be clear about the different concepts:

- **Directory**: is the current organizational structure.
- **Current Directory**: is where I am (pwd) ".".
- **Root Directory**: system mount point /, there is only one per system.
- **Father Directory**: it is the previous directory to the current one, its existence or its reference is identified by means of ".. with Current Directory".

Linux directories

DIRECTORY	DESCRIPTION
/bin/	Commands / essential binary programs (cp, mv, ls, rm, etc.).
/boot/	Files used during system startup (core and RAM disks) /dev / Essential devices, hard disks, terminals, sound, video, dvd / cd players, etc.
/etc/	Configuration files used throughout the system and that are specific to the computer.
/etc/opt/	Configuration files used by programs hosted within /opt / / etc / X11 / Configuration files for the X Window system (Optional).
/etc/sgml/	Configuration files for SGML (Optional).
/etc/xml/	Configuration files for XML (Optional).
/home/	Directories of user starts (Optional).
/lib/	Shared libraries essential for the /bin/, /sbin/binaries and the system kernel.
/mnt/	File systems mounted temporarily.
/media/	Mounting points for media devices such as CD players.
/opt/	Static application packages.
/proc/	Virtual file system that documents events and kernel states. It contains mainly text files.
/root/	Home directory of the root user (super-user) (Optional).
/sbin/	Commands /binary system administration programs.
/tmp/	Temporary files.
/srv/	Site-specific data served by the system.
/usr/	Secondary hierarchy for read-only shared data (Unix system resources). This directory can be shared by multiple computers and must not contain specific data of the computer that shares them.
/usr/bin/	Commands / binary programs.
/usr/include/	Standard inclusion files (header headers used for development).
/usr/lib/	Shared libraries.
/usr/share/	Shared data independent of the architecture of the system. Images, text files, etc.
/usr/src/	Source codes (Optional).
/usr/X11R6/	X Window System, version 11, launch 6 (Optional).
/usr/local/	Tertiary hierarchy for shared read-only data specific to the computer that shares them.
/var/	Variable files, such as logs, databases, root directory of HTTP and FTP servers, mail queues, temporary files, etc.
/var/cache/	Cache gives application data.
/var/crash/	Information repository referring to system crashes (Optional).
/var/games/	Variable data of applications for games (Optional).
/var/lib/	Variable status information. Some servers such as MySQL and PostgreSQL store their databases in subordinate directories of it.
/var/lock/	Blocking files.
/var/log/	Files and directories of system registry (logs).
/var/mail/	User mailboxes (Optional).
/var/opt/	Variable data of /var/opt/. Applications.
/var/spool/	Application data queues.
/var/tmp/	Temporary files preserved between reboots.

STEP 1: Display the current directory.

        pwd

a)   Help.

        pwd --help

b)   Default value.

        pwd

STEP 2: Access a directory.

        cd

a)   Help

a.1) Online command help.

     cd --help

a.2) Help with applications.

     man   cd

     info   cd

b)   Access a directory, as input.

     cd   directory

     cd   directory path

   ej.: cd /etc/network

c)   Exit a directory.

  c.1) Access to the previous directory.

     cd ..

  c.2) Access to the root directory.

     cd /

  c.3) Access to the HOME directory, it is the working directory (home), the environment variable contains that route, $ HOME, it is the user's directory. The next three lines access the home directory of the active use.

     cd

     cd ~

     cd $HOME

   Access the / boot directory and list its contents.

     cd boot

     ls -l

STEP 3: View partitions and filesystems and disks.

fdisk	Partition and visualize.
df	View mounting point information.
mount	Mounted file systems.
/dev/sda1	/
/dev/sda2	/swap
/dev/sda3	/mnt/local

a)   View fdisk help.

     fdisk --help

     man fdisk

     info fdisk

b)   Display df help.

     df --help

     man df

     info df

c)   View help mount.

     mount --help

     man mount

     info mount

     infotext   info

     textinfo

   Example:

     fdisk –l

     df

     mount

> **DAEMON** (processes or tasks resident in memory, in the background).
> The word demon comes from the acronym in English D.A.E.M.O.N (Disk And Execution Monitor) which is a special type of computer process that runs in the background instead of being controlled directly by the user (it is a non-interactive process).
> **Characteristics:**
> - They do not have a direct interface with the user, either graphic or textual ...
> - They do not make use of the standard inputs and outputs to communicate errors or register their operation, but use system files in special zones (/ var / log /) or use other specialized daemons in that registry as the syslogd.

STEP 4: Visualize the tree structure.

     tree

a)   Help.

     tree --help

b)   Default.

     tree

c)   List the directories only.

     tree -d

     dir

     ls -l

     alias

> **dir:** is a CMD | COMMAND command and works on Linux, if it is already defined by means of an alias.
> **alias:** abbreviations, redefinitions of orders to streamline work, by others known in other systems.
> Define: # alias dir='ls -l'
> Display alias;   # alias

STEP 5: Clear the screen.

     clear

a)   Help.

     man clear

```
      clear --help
b)  Default.
      clear
```

STEP 6: Create directories.

mkdir

a) Help.
```
      mkdir --help
      man mkdir
      info mkdir
```
b) Create directories.
```
      cd   /mnt/local
      ls -l
```
b.1) There is something if lost + found (it is created by each file system and new mount point).
```
      cd lost+found
      ls -l              It does not have visible information, initially.
      ls -la             Display hidden information.
      ./                 Current Directory.
      ../                Reference to the previous directory (parent).
      cd ..              Exit to the previous directory.
      ls -l              Display the contents of the current directory.
```
c) Create directory and visualize the tree structure.
```
      mkdir fray
      mkdir diego
      mkdir estudy
      mkdir others
      tree
```

> If tree does not work is that it is not installed you have to load the package (Debian, ...)
> **apt-get install tree**

c.1) Visualize tree structure in graphic format.
```
      tree -A
```
d) Create more than one directory simultaneously.
```
      mkdir toros futbol deportes
```
e) Create a complex directory structure. The option [-p] is used to create all the parent structures of the final directory, they are created in the same line of execution
```
      mkdir -p hacienda/declara/pillan
      mkdir -p hacienda/fraude/trincan hacienda/recaudación
```

STEP 7: Delete directories.

It allows to delete files and directories directly or recursively.
```
      rm
```
a) Help.
```
      rm --help
      man rm
      info rm
```
b) Delete by default (files).
```
      rm futbol
```
Delete the directory as long as it does not have files or other directories.
Default Delete files.
c) Delete directories by default.
```
      rm -r futbol
```
r erase recursively.
d) Delete directory structure with files.
```
      rm -r hacienda
```
e) Clear force (brute force).
```
      rm -rf hacienda
```

rm	
Syntax:	**rm [option] file**
OPTION	DESCRIPTION
-f	Ignore files were conspicuous by their absence, and will never ask before removing.
-i	Ask before each extraction.
-I	Ask once before removing more than three files, or when withdrawing recursively. Less intrusive than -i, while still protecting against most errors.
-r , -R	Delete directories and their contents recursively.

STEP 8: Clear force (brute force).

The touch command allows you to create the entries in the table of inodes, but the file does not contain information, it does not occupy any of the direct and indirect addressing entries.
```
      touch
```
a) Help.
```
      touch --help
```
b) Defect. We create a file whose size is empty.
```
      touch negro
      ls -l
      cd ..
```

ping 192.168.0.100
We test if the IP and the gateway work, for this we access with a text browser to the internet.
lynx http://www.google.es
c) Update the date and time of a file that already exists.
ls -l /etc > prove000
ls -l
touch prove000
ls -l

STEP 9: Move a directory.

The mv command is the abbreviation of move. It is used to move/rename a file from one directory to another. The mv command completely removes the file from the source and moves it to the specified folder.
mv
a) Help.
mv --help
man mv
info mv
mount
df
cd /mnt/local
b) See the version of the order.
mv -v
mv --verbose
mount -v
df -v
c) Move by default.
mv others toros
tree
d) Change of name.
mv toros vacas
e) Change the forcing name.
mv -f toros vacas
f) Interact during the change.
mv -i toros vacas

mv
Syntax: mv [*option*]... source... *directory*

OPTION	DESCRIPTION
-b	As --backup but do not accept any argument.
-f	Never ask before overwriting.
-i	Ask for confirmation before overwriting.
-S	Replace the usual backup suffix.
-T	Treat DESTINATION as a normal file.
-u	You only move when the ORIGIN file is more modern than the destination file.

PRACTICE 7: Handle the most common apt-get options
DESCRIPCIÓN:

The apt is a very powerful and easy to use tool, we can forget having to use sources, compile, if libraries, if I have to install such rpm, if I need a newer one than the one that comes in the CD of the distribution, now nothing, always apt, luckily, 99.44% of the software for Linux is "debianizado", that is, it is precompiled and ready to install it in your Debian. That's why DEBIAN IS THE BEST, and the easiest to use.

Installation of packages.

apt-get install nombre_paquete1 pakete2 paquete3

Search for packages.

apt-cache search texto_a_buscar

Update System.

apt-get update

apt-get upgrade

EYE! First of all you have to have the **/etc/apt/sources.list** file properly configured.

REQUISITOS:

Installation of packages other than those requested by default.

apt-get install paquete/unstable

apt-get install paquete/testing

In general, packages of the stable type are usually obtained by default, but these usually have versions of older programs, so we may want to have more recent packages, such as testing types. For more information, see the Configuration Files section.

a) Reconfigure a package.

dpkg-reconfigure nombrepkt

This can be useful for example to reconfigure the X or local, I've also used it once with the etherconf or with iptables to tell you to load them when the computer starts up. Example:

dpkg-reconfigure iptables

dpkg-reconfigure locales

dpkg-reconfigure etherconf

Deleting installed packages.

apt-get remove nombre_pkt

b) Deleting installed packages.

Example of a source file: **sources.list**

> #las líneas que comienzan por # son comentarios.
> #Actualizaciones de seguridad! Básicas y necesarias!
> deb http://security.debian.org/ stable/updates main
>
> deb ftp://ftp.es.debian.org/debian stable main contrib non-free
> deb ftp://http.us.debian.org/debian stable main contrib non-free
>
> #Paquetes testing
> deb http://ftp.rediris.es/debian/ testing main contrib non-free
> deb http://ftp.rediris.es/debian-non-US/ testing/non-US main contrib non-free
>
> # Paquetes Inestables
> deb http://ftp.es.debian.org/debian/ unstable main contrib non-free
> deb http://ftp.es.debian.org/debian-non-US/ unstable/non-US main contrib non-free
> deb http://ftp.rediris.es/debian/ unstable main contrib non-free
> deb http://ftp.rediris.es/debian-non-US/ unstable/non-US main contrib non-free

An interesting program is the netselect used to find the list of closest sources that work best.

netselect-apt tipo_paquete

Where type of package is: **stable, unstable o testing.**

/etc/apt/apt.conf.d/70debconf

By default stable packages are installed, which are very tested and which in principle do not have any type of dependency conflicts, however it is also true that they are usually old versions of software, and since many programs are in continuous development maybe we interest to have more recent versions with better features, and even paradoxically more stable to be versions with fewer errors. To do this we just have to add APT :: Default-release "package_type" where package type is stable, testing or unstable. The testing versions in my opinion are the most comf.

cat /etc/apt/apt.conf.d/70debconf

> // Pre-configure all packages with debconf before they are installed.
> // If you don't like it, comment it out.
> DPkg::Pre-Install-Pkgs {"/usr/sbin/dpkg-preconfigure—apt || true";};
> APT::Default-Release "stable";

c) Manage apt-get and dpkg.

Some possibilities of the tools:

apt-get y dpkg de Debian GNU/Linux

List all the files in a package:

> **$dpkg -L nombre_paquete**

Install a package of a specific release:

> **# apt-get install -t unstable nombre_paquete**

Block (hold) a package so that it is not updated in the upgrades:

> **# echo nombre_paquete hold | dpkg --set-selections**

Remove the block to a package:

> **# echo nombre_paquete install | dpkg --set-selections**

See the version of an installed package:

> **$ apt-cache policy nombre_paquete | grep Installed**

List the packages that contain a certain string in their name:

> **$ COLUMNS=120 dpkg -l | grep string**

Get the stateB (hold, purge) of a package:

> **$ dpkg --get-selections nombre_paquete**

Delete a package and its configuration files:

> **# dpkg --purge nombre_paquete**

Delete a package and its configuration files:

> **$ apt-cache showpkg nombre_paquete**

Delete a package and its configuration files:

> **$ apt-cache search string**

d) Possible problems.

When installing a package, it may happen that your post-installation script fails for some reason, which prevents the package from installing correctly. If that happens you can edit your corresponding script in:

/var/lib/dpkg/info/name_packet .postinst

and try to fix it. Then simply execute:

> **# dpkg --configure -a**

Reinstall all installed packages.

Useful to clean the binaries if the system has been infected with a virus or a rootkit.

USE WITH CAUTION.

> *# for i in $(dpkg—get-selections | grep -v deinstall | awk '{print $1}'); do apt-get install -y—reinstall$i; done*

WORK UNIT III: Files in Linux

PRACTICE 8: Types of files.

PRACTICE 9: Change or establish permissions and properties.

PRACTICE 10: Handle text files in Linux.

PRACTICE 11: Search of files.

PRACTICE 12: Create and manage devices.

PRACTICE 13: Show files that exist in a Linux structure.

PRACTICE 14: Processing of files in Linux.

PRACTICE 15: Create soft and hard links or links in Linux.

PRACTICE 16: Access the definition of Environment in Linux.

Commands

vi, touch, nano, pico, echo, umask, cat, less, more, rm, mv, chmod, chown, less, pg, wc, head, tail, cut, locate, ls, slocate, whereis, whatis, find, grep, ln, egrep, mount, umount, lsusb, eject, fuser, sort, comm, diff, gzip, gunzip, zcat, zmore, zcmp, zdiff, symlink, cal, ncal, tar, calendar, date, uptime, lwclock, watch, set, env, alias, unalias, uniq, sum, lsof, paste, lshal, biosdecode, lsattr, chattr, dmidecode.

Contents
- Introduction to the files.
- Types of files in Linux.
- Metacharacters.
- Operations with files.
- Permissions for files.
- Attributes of the files.
- Compression of the files.
- Text editing in Linux.

PRACTICE 8: Types of files.

DESCRIPTION:

The files are identified by their permissions, there are 10 characters that identify them: The first character the type of file (-) is identifying the file, the characters from 2 to 4 identify the permissions of the owner of the file, from character 5 to 7 identify the permissions of the group to which it belongs and the last three characters identify the permissions to other users and other groups.

        rwxrwxrwx   --> u g o

Permissions to access a file

        r   Reading
        w   writing
        x   executable

Types of permits

        u   - User permissions (the one that created it)
        g   - Group permissions
        o   - Others

When creating users, there are the default permissions, by means of masks.

        r   w   x
        2^2 2^1 2^0 --> Numbering System. (OCTAL)

        8 / 2 = 2^3

        r-x rw- r--   literal expression
        101 110 100   A direct conversion from octal to binary is performed
          5   6   4   Permit value numerically
        u=+rx g=+rw   o=+r -> (ugo) permissions using literals.

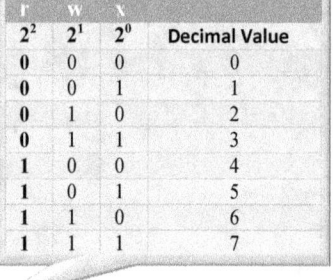

r	w	x	
2^2	2^1	2^0	Decimal Value
0	0	0	0
0	0	1	1
0	1	0	2
0	1	1	3
1	0	0	4
1	0	1	5
1	1	0	6
1	1	1	7

Permissions Mask

umask: is the abbreviation in English of user file creation mode mask (mode of the creation mask of files), that is to say, the format of permissions that are going to have the files and the directories that the user is creating, therefore, this command It is used to set the permissions that the new files and directories created by default have.

STEP 1: Create a text file.

There are different ways to create a text file, from the shell line.
a)   Create an online file of orders ($,#, >).
a.1) Create an empty file.
        touch no*mbre_fichero
a.2) Create a file with address.
        cat > fich001

        .....
        CTRL+D
EJ:
        # cd /mnt/local
        # touch eje001
        # cat > eje002
        Good morning Juan
        Revolcón or big door
        ^D
        # ls -l
b)   Create files using line editors or text editors.
        vi eje004
        nano eje004
        pico eje005
c)   Using a word processor, ...,
        gedit (in graphic environment).
d)   Using a word processor
        echo   Other line > ejer005
        vi ejer005
        nano ejer005
e)   It can be done with any command that allows to visualize information.
        echo   this file is new > eje006
        ls -l
        cat eje006
        echo   add a second line >> eje006
        cat eje006

STEP 2: Permission mask.

To modify the value of umask permanently it will be necessary to include this configuration in /etc/profile or /etc/bash.bash affecting the change to the whole system; or in the files ~/.profile or ~/.bashrc if you want to apply the change for a specific user.

The umask command makes the difference at the bit level using the AND operator

        umask

a)   Help.

           umask --help

           man   umask

           info   umask

b)   By default, it allows to visualize the active mask.

           umask

The mask corresponds to 4 blocks of 3 characters. It is expressed numerically.

     0 6 2 2-> the first digit corresponds to special or hidden permits.

     u g o --> visible permissions.

c)   Check the permits.

           umask -pS

d)   Assign mask permissions in octal.

           umask   0622

           touch   eje003

           umask -pS

           ls -l

           umask   0022

           umask -pS

e)   Assign mask permissions with literals.

           umask u=rwx,g=rwx,o=

           touch   ejer004

           umask -pS

STEP 3: Display the contents of a plain text file.

Visualize with cat and equivalents

        cat

        more

        less

a)   Visualize with cat.

a.1) Help.

           cat --help

a.2) Default.

           cat   eje006

a.3) Display the contents of a file, numbering the lines.

           cat -b eje006

b)   Display with more.

Show the contents of a text file.

b.1) Help.

           more --help

b.2) Display in SCROLL, screen to screen.

           more eje006

b.3) Do not use the scroll.

           more   -p eje006

           more   /etc/passwd >>eje007

           more   /etc/passwd >>eje007

           more   /etc/passwd >>eje007

           more   /etc/passwd

           more   -p /etc/passwd

           more eje007

           more -p eje007

c)   Visualize within the program, with mobility.

c.1) Help.

           less --help

c.2) Visualize by default.

           less eje007

           (press q -> quit to exit)

d)   Display with addresses, input.

           cat < eje007

           more <eje007

           less   <eje007

---

umask

Calculation of the final permit for FILES.

You can simply subtract the umask from the base permissions to determine the final permission for the file as follows:

       666-022 = 644

- File base permissions: 666
- Umask value: 022
- Subtract to get new file permissions (666-022): 644 (rw-r- r--)

Calculation of the final permission for the directories.

You can simply subtract the umask from the base permissions to determine the final permission for the directory as follows:

       777-022 = 755

- Directory permits Directory : 777
- umask value: 022
- Subtract to get new directory permissions (777-022): 755 (rwxr-xr-x)

---

cat

Syntax: cat [-s] [-v[et]] [arhive ...]

OPTION	DESCRIPTION
-v	Displays control characters (not printable).
-s	Replace several blank lines with a single line4
-t	Like -v but also print tabs like ^ I
-e	The same as -v but it also prints $ at the end of each line.

---

more

Syntax: more [-dpcsu] [-num] [+/Pattern] [+linenum] file ...

OPTION	DESCRIPTION
-num	Specify the screen size (in lines)
-d	Display the message "[Press space to continue, 'q' to quit]".
-p	It does not scroll, but it cleans the screen and displays the text.
-c	It does not "scroll", it displays line by line from top to bottom.
-s	Replace several consecutive blank lines with one.
-u S	Delete underlined
+/Pattern	Start with the page that contains the word pattern.
+linenum	Start in the line linenum.

---

less

Syntax: less [options] list_of_files

OPTION	DESCRIPTION
-e	It causes less to exit automatically the second time it reaches the end of the file. By default, the only way out from less is the order q.
-E	It causes less to exit automatically the first time it reaches the end of the file.
-n	Deletes line numbers, and replaces them with the byte number where the line is in the whole file.
-Q	Suppress all acoustic signal in the search.
-s	It causes several consecutive blank lines to be compressed into one.

STEP 4: Change the name to a file or move it to another directory.

mv

a)   Help.

mv --help

man mv

info mv

b)   Change the name of a file.

mv eje007   eje008

ls -l

c)   Move a file to a directory.

mv eje008 vacas

tree

mv vacas/eje008   .   (He . "Pto" refers to the current directory)

mv		
Syntax:	**mv [-options] file1 file2 ..... directory**	
OPTION	**DESCRIPTION**	
-i	Ask for confirmation before overwriting	
-f	It does not ask for confirmation.	
-b	Create backup copies of files that are going to be overwritten or deleted.	
-u	It does not move a file or directory that has an existing destination with the same modification time or more recent.	

STEP 5: Copy files.

cp

a)   Help.

cp --help

NOTE: caracteres comodín *, ?

b)   Copy a file.

cp eje* vacas

tree

c)   Duplicate a file.

cp eje008   eje007

d)   Force to delete a file (FICH-DIRECTORIOS).

cp -f eje* estudias

e)   Delete recursively.

cp -r .   fray

cp -r /etc fray

f)   Visualize what is being copied.

cp -r -v /etc fray

tree

tree –d

cp	
Syntax:	**cp [options] file_source file_target**
OPTION	**DESCRIPTION**
-b	Create a backup in the destination in the case where there is a file called the same as the one we want to generate.
-f	Force the deletion of the destination files without consulting or notifying the user.
-i	Report before overwriting a file in the indicated destination.
-l	Make a link instead of copying the files.
-p	It makes the copy of the files and directories keeping the date of modification of the original files and folders.
-r	Copy recursively.
-S SUFFIX	Add the word "SUFFIX" (or the word we indicate, for example BACKUP) to the backup files created with the "-b" flag.
-u	The cp command in Linux does not copy a file or directory to a destination if this destination has the same modification date or a later modification date comparing it with the file or directory that we want to move.
-v	It shows what is running.

STEP 6: Delete a file or directory.

rm

a)   Help.

rm --help

b)   By default, delete, without entering directories.

rm eje008

c)   Delete recursively.

rm -r fray

d)   Clear force.

rm -f fray

e)   Ask before deleting (interactive form).

rm -i fray

f)   Clear forcefully and recursively.

rm -fr fray

rm -f -r -i fray

rm -ri vacas

tree

rm	
Syntax:	**rm [-if] file1 [file2 ...]**
	rm [-ifrR] directory1 [directory2 ...]
OPTION	**DESCRIPTION**
-i	Interactive (ask for confirmation)
-f	It does not issue error messages when the file or directory does not exist.
-r-R	Recursive Delete a directory and all its contents.

PRACTICE 9: Change or establish permissions and properties.
DESCRIPTION:

Who can be granted permits?
The permits can only be granted to three types or groups of users:
- To the user who owns the file.
- To the group that owns the file.
- To the rest of the users of the system (all except the owner).

Permissions:
They are displayed with ls -l

drwxrwxr--   We visualize permissions and identification of files.

The first character is identifying the type of file system.

> **d** -> directory identification.
> **c** -> character (device and communication mode is at character level.
> **b** -> block (device and communication mode at block level.
> **p** -> pipe (pipeline or filter).
> **s** -> socket.
> **l** -> link level.

Files can have the permissions of:

> **r**   reading
> **w**   writing
> **x**   executables

A file according to the storage format can be:
- They are binary files -> created by compiler: gcc, cc, ...).
- Text files -> SCRIPT (script file, depending on the shell type: sh, bash, csh, tcsh, kcsk, zcsh, zsh, ...).

Nomenclatura

> Numeric (OCTAL) 734   rwx –wx r—
> Literals: u g o (+ |-) rwx
>   a

There are special permits.
They are usually specified with the umask mask.

Change the owner and the group
To change the owner user and the owner group of a file or folder, use the chown (change owner) command. To do this you must have write permissions on the file or folder. The syntax of the command is:

> # chown new_user [.new_group] file_name XZ

There are 3 types of users:
1. Normal User: is a particular individual that can enter the system, with more or less privileges that will make use of the resources of the system. As an indicator at the prompt, use the symbol $ (dollar). Example: raul, sergio, mrodriguez, etc. They are also known as login users.
2. System users, are own users of the system linked to the tasks that the operating system must perform, this type of user can not enter the system with a normal login. Example: mail, ftp, bin, sys, proxy, etc. It is also known as users without login.
3. root (superuser), all operating system GNU / Linux has a superuser, which has the maximum privileges that allow you to perform any operation on the system, its existence is essential as it is responsible for managing servers, gro

STEP 1: Order to change the permissions.
The chmod (change mode) command is the command used to change permissions, you can add or remove permissions to one or more files with + (more) or - (less).

> chmod

a)   Help.
> man chmod
> info chmod

> chmod --help

b)   Change the permissions numerically.
> chmod   permisos   [file|directory]

> The permissions are three numbers: 777   000
> chmod   770 eje001

c)   Create a file.
> cat >eje010
> *!#/bin/bash*
> *clear*
> *echo   Good Mornin*
> *^D*
> umask
> 0022   --> 110 100 100

For a script to be executable you have to change the permissions:

```
chmod  744  eje010
rwxr—r--    eje010
```

STEP 2: Execute a file.

a)   If it is a script, it can be preceded by the Shell that you want to execute it.

```
sh  eje010
bash  eje010
csh  eje010
```

or any other program.
Precede the file of the current route

b)   It is the identification of the current directory.

```
name_file
```

Ej.:   .   eje010

```
./name_file
```

Ej.:   ./eje010

> The ending of the file name and appears:
> /   indicates that it is a directory.
> *   indicates that it is an executable file.

c)   It can be executed directly, if the search path of the operating system contemplates the current path of the executable file.

```
cp
mv
```

You can find the search path for executable files, in an environment variable.

```
PATH
```

Visualize the environment variables.

```
set
cd                        Access the HOME directory
ls -l                     There is zero file
ls -la                    View hidden files in long format
less .bash_history        File that contains the history of executed orders.
```

d)   The files that the first character is. They are hidden files.

```
cd /mnt/local/deportes
ls –l
```

e)   Hide a file, change its name and put the first character a point (.).

```
mv    eje010   .eje010
```

f)   Display the contents of the directory.

```
ls -l
```

g)   View the contents of the directory and the hidden files.

```
ls -la
```

h)   Descullate a file. The first character is renamed and removed (.)

```
mv  .eje010  eje010
```

i)   Create a hidden file, from the console.

```
cat   >.eje011
!#/bin/bash
echo    second script file
echo    it is a hidden file
^D
```

j)   Change the permissions to a directory.

```
mkdir   alumno
ls -l
chmod   700 alumno
ls -l
```

> Set end of file in a command
> line is: ^ D or CTRL + D

k)   Hide a directory.

```
mv    alumno  .alumno
ls -la
ls –l
```

STEP 3: Change the permissions with literals.

The chmod (change mode) is the command used to change permissions, you can add or remove permissions to one or more files with + (more) or - (less).

```
chmod
```

a)   Change the permissions.

```
chmod  ug-rwx eje010
ls -l
chmod  ug+rw eje010
ls -l
chmod  u+x g-w o+rx eje010 --> incorrecto
chmod  u+x eje010
chmod  g-w ej010
chmod  o+rw  eje010
```

> In order to modify and eliminate the permissions to be useful, one must be able to modify the directory in which the file is located:
> #chmod ugo + rwx ./

The literals are all together or must be executed in independent orders.
b)   Establish all permits to all parties (ugo).
        chmod ugo+rwx eje010
        chmod a+rwx eje010
        chmod a-rwx eje010
        ls -l
        chmod ugo+rw eje010
        chmod ugo-w+x eje010
c)   Grant reading access + r in a file to all members of your group (g).
        chmod g + r   eje010
d)   Grant reading access to a directory to all the members of your group:
        chmod g + rx /practicas
e)   The "execute" permission is required to read a directory.
        chmod ugo +r   /practicas
f)   Grant permissions to read everyone on the system in a file that you own so that everyone can read it: (u) user, (g) group and (or) other.
        chmod ugo + r   ejer010
g)   Grant reading and execution permissions in a system directory.
        chmod ugo + rx ejer010
h)   Grant reading and modification permissions to a file that you own.
        chmod ugo + rw ejer010
i)   Denying read access to a file by everyone except himself.
        chmod go -r ejer010

STEP 4: Change the owner of a file.

The chown command allows you to change the owner of a file or directory on Linux systems. You can specify both the name of a user, as well as the user identifier (UID) and the group identifier (GID). Optionally, using a colon (:), or a period (.), Without spaces between them, then the user and group to which each file belongs is changed.

Each Linux file has an owner and a group, which correspond to the user and the group who created it.

The root user can change the owner of any file or directory.
        chown
a)   Help.
        chown --help
b)   Change the owner of a file. First you put the owner and then the name of the file.
        chown smr eje010

STEP 5: Register a user.

The smr user is registered in the /home/smr directory if the assigned working directory does not exist -m allows the directory to be created.
        useradd -m -d /home/smr   smr
View the user file and note that the last line contains the user name smr (less allows mobility or scrolling in the file, up and down, page forward or back page, to exit press q-> quit).
        less /etc/passwd
Once the smr user is registered, the password or password of this user is established, we use the passwd command.
He asks us for the key and we type Practica2015 *. The keys must be entered twice, the second is validation.
        passwd smr
        : Practica2015*
Open the connection of a new console, ex .: CTRL+ALT+F2.
        login: smr
        password: Practica015*

> *If we do not specify the name of the user to change the key, the active user assumes the default.*

PRACTICE 10: Handle text files in Linux.

DESCRIPTION:

For Linux everything is a file system, there are no units, no devices, everything is treated as a file system, and there is only one root directory.

From the root directory:

- There is only one root directory per operating system boot, and all the rest part of it.
- It is represented by a symbol (\).

Units and devices must be mounted to drive.

A unit is mounted as one more directory, from the point of assembly. Ex:

```
mount /dev/sdb1  /mnt/disco2
cd    /mnt/disco2
ls -l
```

Processing of files.

- Display files.
- Count words.
- View headers of the files.
- See footer of a file.
- Search files.
- Compress /files.
- Copy files.
- Mount /unmount/view file systems.
- Repair file systems.

STEP 1: Display the content of plain text files.

There are different commands that allow viewing the contents of a file whose content is text, or a flat text file. Of all of them the most powerful is less, although all can be used with pipes or pipeline.

```
cat
more
less
pg
```

It allows interactivity, with the editing keys.
q--> quit of less

a) Help.

```
cat --help
more --help
less --help
pg --help
```

b) Pipes, filters, pipe or pipeline.

```
command1 | command2 |command3
more
pg
less
```

Examples:

```
cat  eje007
more eje007
pg   eje007
less eje007
cat  eje007|more
cat  eje007|pg
cat  eje007|less
ls -l /bin  |less
```

c) Addresses and redirects.

```
cat > eje012
cat <eje007
pg  < eje007
more <eje007
less < eje007
```

OPERATOR REDIRECTION	DESCRIPTION
>	Output address, device or a file (if it exists I charge it and I create it again)
<	Entry Addressing.
>>	Redirection of output, device or file (if it does not exist, it creates the file and exists, it adds it to the end).
>&	Output Addressing.
<&	Input Addressing.
|	Pipe, filter or pipeline, the output that performs the order from the left in the standard device is picked up and sent to the order to the right: **command | command**

Handle	NUMBER EQUIVALENT TO THE HANDLE	DESCRIPTION
STDIN	0	Input by keyboard
STDOUT	1	Output of the command to the prompt.
STDERR	2	Output error of a command, displayed at the prompt.
UNDEFINED	3-9	Undefined, waiting for defined individually by the application and are specific to each tool.

STEP 2: Count words from a text file.

```
wc
```

a) Help.

```
wc --help
```

b) By default.

```
wc eje007
```

c) Count lines.

```
wc -l eje007
```

d) Count words.

```
wc -w eje007
```

e) Count the bytes.

```
wc -c eje007
```

WC	
Syntax:	wc [-options] list_file
OPTION	DESCRIPTION
-c	Displays the number of characters. Specifically, it counts the number of line returns.
-L	Displays the length of the longest line.
-l	Displays only the number of lines
-m	Displays the number of characters.
-w	Displays the number of words. A word is a string separated by a space, a tabulator or a new line.

f) Count the characters.
      wc -m eje007

STEP 3: Display the header or initial part of a text file.
      head
a) Help.
        head --help
        man head
b) By default.
        head   eje007
    Displays the first 10 lines of the file.
c) Display a specific number of lines (whichever is specified).
        head -4 eje007
        head -15 eje007
        head -22 eje007
d) Display a specific number of bytes.
        head -c256 eje007
        head -c643 eje007
e) Version of the command.
        head --versión
        head -v
f) Display the headers of several files.
        head eje*
        head -v -c110 eje*
g) Visualize several files the content of headers.
        head eje* prac*
        head -v -c150 eje* prac* pru[1-3]*

head
Syntax: head [options] name_of_file

OPTION	DESCRIPTION
-n	Specify how many lines you want to show.
-n number	The number must be a decimal integer whose sign affects the location in the file, measured in lines.
-c number	The number must be a decimal integer whose sign affects the location in the file, measured in octets.

Wildcard characters

*	replaced by one or more characters
?	replaced by a character.
[123678]	
[1-3,6-8]	Ranges of a character
.	Current directory
..	Previous directory
~	Home
~ user	User's home
&	Background
<Ctrl> Z	For a process
;	Separate commands
\	Continuous command line
!!	Repeat the previous command
! n	Execute the command number "n"

STEP 4: Footer of a text file.
  Display the last lines of a text file, by default the last 10 lines.
      tail
a) Help.
        tail --help
b) Default display, only displays the last 10 lines.
        tail eje007
c) Display the name of the file.
        tail -v eje007
d) Display a specific number of lines.
        tail -5 -v eje007   incompatible, these two options together.
        tail -2 -v eje*   incompatible, these two options together.
        tail -5 eje007   correct.
        tail -2 eje*   correct.
        tail -n5 -v eje007   correct.
        tail -n2 -v eje*   correct.
e) Display the last xxx bytes
        tail -c210 eje*
f) Example to extract from a file from line 5 to line 10.
        head -10 eje007 |tail –n6

tail
Syntax: tail [options] name_of_file

OPTIONS	DESCRIPTION
-l	Specifies the line units.
-b	Specify the block units.
-n	Specify how many lines you want to show.
-c number	The number must be a decimal integer whose sign affects the location in the file, measured in bytes.
-n number	The number must be a decimal integer whose sign affects the location in the file, measured in lines.

NOTE: tail head, cut They are used in pipes

Example to extract from a file from line 5 to line 10.
  It allows selecting by a column whenever they are established with tabs or delimiters.
      cut
a) Help.
        cut --help
b) Create a DB file, with limiter.
        nano toros
c) Field delimiter.
        cut -d: toros
        cut -d: -f2 toros   only 2nd field
        cut -d: -f3 toros   only 3rd field
        cut -d: -f1,3 toros
        cut -d: -f1,3-5 toros   the 1st field, the 3rd until the 5th
        cut -d: -f1 /etc/password
        cut -d: -f1 /etc/password > users
    Show a number of characters.
        cat toros
        cut -c20 toros

cut
Syntax: cut [options]

OPTION	DESCRIPTION
-c	Specifies the positions of the characters.
-b	Specifies the positions of the octets.
-d flags	Specifies the delimiters and fields

```
        cut  -c20-45 toros
```
d) Show a byte.
```
        cut  -b3 toros
        cut  -b3-8 toros
```

   Filter adjacent lines that match the INPUT (or the standard input), by typing in OUTPUT (or standard output). If no option is given, the matching lines are combined in the first occurrence.
```
        uniq
```
a) Help.
```
        uniq --help
```
b) Display by default.
```
        ls  -l  >salida010
        ls  -l  >>salida010
        uniq  salida010
        total 8
        rw-------. 1 root root 1234 ago   6 01:55 anaconda-ks.cfg
        lrwxrwxrwx. 1 root root    4 ago   9 22:23 dos0 -> uno0
        lrwxrwxrwx. 1 root root    4 ago   9 22:23 dos1 -> uno1lrwxrwxrwx.
        rw-------. 1 root root 1234 ago   6 01:55 anaconda-ks.cfg
        less  salida010
        total 8
                  rw-------. 1 root root 1234 ago   6 01:55 anaconda-ks.cfg
        lrwxrwxrwx. 1 root root    4 ago   9 22:23 dos0 -> uno0
        lrwxrwxrwx. 1 root root    4 ago   9 22:23 dos0 -> uno0
        lrwxrwxrwx. 1 root root    4 ago   9 22:23 dos0 -> uno0
```

   It shows the checksum and the number of blocks for each FILE.
```
        sum
```
c) Help.
```
        sum --help
```
d) Sum of verification by default.
```
        sum   salida
        08294   1
```
e) Use the System V algorithm.
```
        [root@localhost ~]# sum -s output
        29528 1 output
```
f) Use the BSD algorithm.
```
        [root@localhost ~]# sum -r output
        08294   1
```

sum	
Syntax:	**sum [OPTION]... [FILE]...**
OPTION	**DESCRIPTION**
-r	Use the BSD sum algorithm, with 1K blocks.
-s, --sysv	Use the System V sum algorithm, with blocks of 512 bytes.

   It is used to paste the content from one file to another. It is also used to set the column format of each line.
```
        paste
```
a) Help.
```
        paste --help
```
b) By default. The result is deposited in the first file.
```
        cat  > fich001
        value1  value2  value3
        value4  value5  value6
        ^D
        cat  > fich002
        valor01  valor02  valor03
        valor04  valor05  valor06
        ^D
        root@192:~# paste fich001  fich002
        value1    value2    value3      Valor01   valor02   valor03
        value4    value5    value6      Valor04   valor05   valor06
```

paste	
Syntax:	**paste [OPTION]... [FILE]...**
OPTION	**DESCRIPTION**
-s	Paste one file after another instead of in parallel.
-d	Reuse characters from the list instead of tabs

c) Using delimiters and a list of delimiters. (spaces = TAB and,)
```
        root@192:~# cat fich003
        primero segundo tercero
        cuarto  quinto  sexto
        root@192:~# cat fich004
        uno     dos     tres,cuatro,cinco
        seis    siete   ocho,nueve,diez
        root@192:~# paste -d fich003 fich004
        uno     dos     tres,cuatro,cinco
        seis    siete   ocho,nueve,diez
```
d) Paste in series.
```
        root@192:~# paste -s fich003 fich004
        primero segundo tercero cuarto  quinto  sexto
        uno     dos     tres,cuatro,cinco        seis    siete   ocho,nueve,diez
```

PRACTICE 11: Search of files.

DESCRIPTION:

Searches with find, this command can be a bit slow when you need to search in a very large directory tree. Here the locate command can help. This does not really look directly for a file in the file system. Search in a database.

Finding files by content (searching for text strings in files).

The standard utilities to search for text strings in files are grep/egrep for the search of regular expressions and fgrep to search for literal strings.

      find, grep, egrep, fgrep

There are support orders to help, to indicate where a file is located.

      locate, whatis, whereis.

There are of course other search commands like awk, sed and grep but they are more focused on searching "inside" the files.

STEP 1: Location of orders.

      locate, slocate, mlocate, rlocate, whatis, whereis.

a) Help.
    locate

b) By default, locate in the database, information about the command to search.
    locate   ls
    slocate   ls

b.1) Install the database.
    updatedb
    /var/lib/slocate/slocate.db

b.2) Search the routes, in which this order appears
    whereis   ls

c) Locate help information about a command.
    whatis   ls   --> manual de ayuda
    cd   /usr/man
    ls -l |pg

d) Summary information of what a command does.
    whatis   ls
    whatis   pwd
    whatis   whereis
    whatis   iptabl es

e) Information of where a file is stored.
    whereis   ls
    ls: /bin/ls /usr/share/man/man1/ls.1.gz
    whereis   pwd
    whereis   man
    whereis   su

f) Information which of the executable file.
    which   ls
    /bin/lsw

g) Location information in the help file of the updatebd database.
    locate   ls

STEP 2: Search for files.
    find

a) Help.
    find   --help
    man   find

b) Define what I want to search or search by name.
    find (ruta) modificadores o argumentos.
    find / -name ¿Qué?
    find / -name ls

c) Search files by size.
    find /dev -size 100k   --> tamaño exacto
    find /dev -size +100k   --> tamaño superior a 100k
    find /dev -size -100k   --> tamaño inferior a 100k

d) Search by users.
    find / -user nombre
    find / -user smr

e) Logical operators -or (OR).
    find / -user alumno1 -o -user alumno2
    find / -user alumno1 -o -user alumno2 -o -user smr

f) Search all files that do not belong to a specific owner or user.
    not   negación
    find / -not -user smr

The command ls displays in a large format the contents of the current directory and passes it through a pipe to the order that displays them screen to screen and the page information appears at the bottom.

whatis
Syntax: whatis [Options] WORD KEY ..

OPTION	DESCRIPTION
-d	Issue debug messages
-w	Keywords contain wildcards
- l	Do not trim the output to the width of the terminal
-r	Interpret each keyword as a record expression.
-v	Allow debugging messages

whereis
Syntax: whereis [options] file

OPTION	DESCRIPTION
-f	Define the search
-b	Search only in binary
-m	Search only manual routes.
-s	Search only original routes

EXCEPTION: Literals are used, but only preceded by (-).

You can cancel the execution of an order by pressing ^ C.
find before searching by users check the existence of that user in **/etc/passwd**.

You can cancel the execution of an order by pressing ^ C.

```
        find  /  -not  -user  root
```

g) Search for files that do not belong to any owner.

```
        find  /  -nouser
        useradd  -m –d /home/alumno1  alumno1
        useradd  -m –d /home/alumno2  alumno2
        passwd  alumno1
```
clave: alumno1 (repetirlo 2 veces)
```
        passwd  alumno2
```
clave: alumno2 (repetirlo 2 veces)
```
        find  /  -not –user root   -o –not –user  alumno1 –o -nouser
        find  /  -not –user root   -a –not –user  alumno1 –a -nouser
        find  /  -not –user root   -a –not –user  alumno1 –a  -not -nouser
```

Starting with the 2.6 kernel version of Linux, each
time a user is created, a group with the same
user name is created.

alumno1 --> alumno1

alumno2 --> alumno2

smr --> smr

h) Search files by groups.

```
        find  /  -group  users
        find  /  -not -group  root
```

h.1) We check the existence of the users and groups created, in the previous point f).

```
        less  /etc/group
        find  /  -group smr  -o  -group  alumno1
```

i) Search by file type.

```
        find  /  -type  tipo
        find  /  -type  d
        find  /  -type  f
        find  /  -type  b
        find  /  -type  c
        find  /  -type  s
        find  /  -type  p
        find  /  -type  l
        find  /  -type  f  -user alumno1   -size +100k
```

TYPE	DESCRIPTION OF THE FILE
f	File
d	Directory
b	Block
c	Character
s	Socket
p	Pipeline
l	Symbolic link

j) Search by permissions.

```
        find  /  -perm  770 -type  f  -user root
        find  /  -perm  700 -type  f  -user root
        find  $HOME  -mtime 0
        find  .  -perm 664
        find  .  -perm  -220
        find  .  -perm  -g+w,u+w
        find  .  -perm  /220
        find  .  -perm  /220
        find  .  -perm  /u+w,g+w
        find  .  -perm  /u=w,g=w
        find  .  -perm  -444   -perm  /222  !  -perm /111
        find  .  -perm  -a+r   -perm  /a+w  !  -perm /a+x
```

PERMITS	NUMERICAL VALUE
rwx	7
rw-	6
r-x	5
r--	4
-rx	3
-r-	2
--x	1

Permits
-number   -symbol
/number   /symbol

k) Search file modified by seniority.

k.1) Search the files for the last access.

```
        find  /  -atime  +4
```
The file has been accessed in the last 4 days.

k.2) Search when a previous modification to:

```
        find  /  -mtime  +4
        find  /  -mtime  -4
        find  /  -mtime  4
```

Assuming that the current date is 21-03-2014
+4 is considered between days: 18 and 21
-4 search by date before or equal to 18
4 search on a specific date (4 days before) on 18

k.3) Search by dates the realization of the last file change.

```
        find  /  -ctime  -5
        find  /  -ctime  +5
        find  /  -ctime  5
```

l) Execute an order with find.

```
        find  /  -type  d  -exec   echo  Directorio = {} \;
        find  /  -type  d  -exec   ls -l {} \;
        find  .  -type  f  -exec   file '{}' \;
```

The command to execute with find must always end in {} \;
{} Indicative of receipt of each entry as a parameter of
the search.
\; Completion, the two characters must go together,
without spaces.

STEP 3: Comparisons with -and, -or and -not.

The find command also includes Boolean operators which makes it an even more useful tool:

```
        find  /  -name 'ventas*'   -and    -mmin 120
        find  /  -name 'Ejer*'     -not    -user ana
        find  /  -iname '*enero*'  -or     -group smr
```

STEP 4: Search information within a file.

It allows to search for information contents within a file.

```
        Grep
```

The file must be in plain text format. (TEXT).
It is used to search in:
- Shell Script or Script Files.
- Files exported from BBDD, in the format of:
  - Columns.
  - Separators; ,: "".
  - Used with pipes |.
  - It is combined with cut, find.

a) Help.
    man   grep
    info   grep
    grep --help

b) Search for a text string (by default).
b.1) grep "string to search" file name.
b.2) find / -name passwd | cut -d : -f1|grep "alumno1".
Ej.:   find / -name passwd | cut -d : -f1|grep "alumno1"
    grep "alumno1" /etc/passwd
    find / -name passwd |grep "alumno1"

c) Search in more than one file.
    cd /mnt/local
    ls -l
    nano eje010
    cat eje010
    cp eje010 eje011
    grep "esta" eje010
    grep "esta" eje01?
    grep "esta" eje01*
    grep "esta" eje?1*

d) Find a chain without considering or considering the difference between uppercase and lowercase.
    grep -i "esta" eje01*
    grep -i "uscu" eje01*

e) Search for complete words (that are not part of a word), that is not a substring.
    grep -iw "uscu" ej010
    grep -i "uscu" eje010
    grep -i "de" eje010
    grep -iw "de" eje010
    grep -i "es" eje010
    grep -iw "es" eje010

f) Find a line number after finding the word that matches the string.
    grep -A2 "ESTA" eje010
    grep -iA2 "ESTA" eje010

g) Search in the lines before the matching word. Show the two lines before the coincident.
    grep -B2 "dos" eje010
    grep -iB3 "dos" eje010

h) Search for N Previous and subsequent lines to the coincident.
    grep -C1 "primera" eje010

i) Search all files recursively.
Paths and wildcard characters are used.
    grep -r -i "root" /etc/*
    grep -r -i -w "root" /etc/*
    grep -riw "root" /etc/*

j) Search for words that do not match the string to be searched.
    grep -v "ESTA" eje010
    grep -v "es" eje010
    grep -vi "es" eje010
    grep -viw "es" eje010

k) Count the number of matches, which appear.
    grep -c "esta" eje010
    grep -ci "es" eje010
    grep -ciw "es" eje010

l) Display only the string to search.
    grep -o "ESTA" eje010
    grep -co "ESTA" eje010 --> It is inconsistent only runs c

m) Display the line number in which the content is found.
    grep -n "ESTA" eje010
    grep -no "ESTA" eje010
    grep -nov "ESTA" eje010
    grep -nv "ESTA" e je010

List of model search patterns that you can most frequently use with grep.

CARACTER	CONCERNS
^	The beginning of a line of text.
$	The end of a line of text.
.	Any unique character.
[...]	Any unique character in the list or range in parentheses.
[^...]	Any character that is not in the list or range.
*	Zero or more occurrences of the preceding character or regular expression.
.*	Zero or more occurrences of any unique character.
\	Ignore the special meaning of the next character.

- Double quotes ("") are used to delimit the text you want to be interpreted as a word.
- Single quotes (') can be used to group sentences with multiple words forming single units, or to make sure that certain characters such as $ are interpreted literally. If you want to write characters like &! $? . ; and \ preceded by a backslash, to be interpreted as normal typographical characters.

In Linux there may be files
Hola.Juan.mio.es
All configurable programs have .conf files

n) Display only the files that contain the word to search.

> grep -l "root" /etc/*
>
> The files that contain the string .conf are the configuration files.
>
> Hello.Juan.mio.es

o) Show the position in the file, of a string.

> grep -b "root" /etc/*

Visualize the files in long format and the output send it to the pipe to be processed by the grep command and extract only those that the first character is a d: Directory.

> ls -l | grep '^d'

Extract from the user definition file (passwd), all the lines defined in the first field or start with a user plus a number between 0 and 9 ex: user1: ....

> grep '^user[0-9]' /etc/passwd

It extracts from file file.txt, all the lines that start with an uppercase / lowercase letter between the A..z and contain zero or more characters and the last character of the line is a number between 0 and 9.

> grep '^[A-Za-z]*[0-9]$' archivo.txt

Visualize all files, even hidden ones, are passed to the output device in a pipeline, which yields the data as input to the grep command, and searches.

> ls –a | grep '^\.[^.]'

Search for a string within the file.txt that:

> grep '^.*,[0-9]\{10,\}' archivo.txt

STEP 5: Use regular expressions with egrep.

> egrep

The grep command understands two different versions of regular expression syntax: "basic" and "extended". The egrep command handles extended format:

egrep is used to search the files of one or more pattern arguments, but uses the regular expression that matches extended (as well as the \ <and \> metacharacters).

a) Help.

> egrep --help

```
egrep <flags> 'regular phrase' <file>
```

STEP 6: Use regular expressions with fgrep.

> fgrep

The fgrep command is used to search the files of one or more pattern arguments. It does not use regular expressions; instead, it does the direct string comparison to find the matching lines of text in the entry.

a) Help.

> fgrep --help

Test knowledge and analyze the results.

> # find / -type f \(-perm -04000 -o -perm -02000 \)
>
> # find / -perm -2 ! -type l -ls
>
> # find / -nouser -o -nogroup -print
>
> # find /home -name .rhosts -print

PRACTICE 12: Create and manage devices.

DESCRIPTION:

Linux systems generally use a static method of device creation, implying that a large number of device nodes are created in /dev (literally, hundreds of nodes) regardless of whether the corresponding hardware device actually exists. This is typically done through a MAKEDEV script, which contains a series of mknod program calls with the major and minor numbers corresponding to every possible device that might exist in the world.

With the use of the Udev method, only the nodes corresponding to those devices detected by the kernel will be created. Because these device nodes will be created every time the system is started, they will be stored in a tmpfs file system (which exists entirely in memory). The device nodes do not need much space, so the memory used is very little.

For security, the mount and dismount operation can only be performed by the superuser. To solve this, you can use the user option in the lines of the /etc/fstab file:

The commands used: mount, umount, mknod, lsusb, eject, fuser.

Device management

Linux uses folders to mount different storage devices.

/etc/mtab	table of mounted file systems, information.
/etc/fstab	table of mounted file systems.

STEP 1: List existing devices.

lsusb

a) Help.

lsusb

b) List the devices by default.

lsusb

c) List the specific devices.

lsusb -e sda

d) List the USB device tree hierarchy.

lsusb

lsusb

Syntax: lsusb [option]...

OPTION	DESCRIPTION
-s [devnum]	Show only devices with specific device and / or bus lines (in decimal).
-d proveedor: [producto]	Show only devices with the specified provider and product identification numbers (in hexadecimal).
-D Dispositivo	Select the device that lsusb will examine.
-t	Dump the physical USB device hierarchy as a tree.

STEP 2: Mount a device.

a) Help.

mount --help

The types can be vfat, ext3, ntfs, xfs, reiserfs, minix and, moreover, to mount a CD-ROM the iso9660 type is used, for an nfs network mount, etc.

On newer Linux systems, it is not necessary to specify the file system because mount automatically detects them.

The dispo and mount_point arguments refer to the device and the directory where the file system is to be mounted. This assembly point must be a directory that exists and that is also empty.

b) Display the assembled units by default.

mount

c) Mount a USB device in the folder /media/usb.

mount -t vfat /dev/sda1 /mnt/usb

d) It allows to mount the image.iso file in / media / cdrom and in read-only mode.

mount -t iso9660 -o ro,loop ~/cd-isodatos.iso /media/cdrom

e) Mount a flash device with the noatime option (this reduces the number of writes).

mount -w -o noatime /dev/sda1 /memstick

f) Mount devices in FAT.

mount -t vfat /dev/sdb1 /mnt/usb

mount -t vfat /dev/hdb1 /mnt/disco1

g) Mount devices in NTFS.

mount -t ntfs-3g /dev/sdb1 /media/usb

mount

Syntax: mount [options] [-t tipo] [-a] [-o opc] device mount_point

OPTION	DESCRIPTION
-a	It allows to mount all the filesystems specified in the file /etc/fstab.
-f	Make a fictitious assembly. It serves to check if the assembly would be carried out correctly.
-n	Mount the device without writing it in the /etc/mtab file.
-r	Mount the file system as read only.
-w	Mount the file system for reading/writing (default option).
-O	Use together with -a, to limit the set of file systems to which it applies. It can be combined with -t and the restriction is cumulative (Slackware's own). mount -a -O no_netdev mount -a -t ext2 -O _netdev
-o	The options are specified with a flag or followed by a string separated by commas of options.

STEP 3: Unmount a device.

It is important to disassemble a disk before removing it. This gives the system the opportunity to complete any pending script and avoids leaving unstable access to device structures by disassembling the file system cleanly.

The use of the umount command guarantees that all information kept in memory by the operating system is written to the device before dismounting it. For this you can also use the sync device command.

umount

a) Help.

umount—help

b) Unmount the floppy disk drive B.

umount /dev/fd1

c) Unmount the device that has been mounted in the / mnt / win directory.

umount /mnt/win

d)   It allows to disassemble all the units.
>   umount -a

e)   Disassemble all mounted systems of the vfat type.
>   umount -t vfat

f)   Unmount a USB device mounted on our computer.
>   umount /dev/sda1
>   umount /dev/sda2

> **umount**
> **Syntax: umount   device | mount_point**
> To disassemble a system, you can use, without distinction, the device or the directory in which it is mounted.

STEP 4: Create devices.
>   mknod

a)   Help.
>   mknod --help

b)   To create a device, we will make a new block device node and call it a floppy disk in the / dev and use the same major and minor numbers of the device.
>   /dev/fd0
>   # mknod /dev/disquete b 1 112

c)   Create a second entry and call raw.diskette, which will be a character device based on the existing device /dev/fd0.
>   # mknod /dev/raw.disquete e 1 112

STEP 5: Display the BIOS devices.
>   biosdecode

a)   Help.
>   biosdecode -help

b)   Default display.

```
root@puesto000:~# biosdecode
# biosdecode 2.11
ACPI 2.0 present.
OEM Identifier: VBOX
RSD Table 32-bit Address: 0x3FFF0000
XSD Table 64-bit Address: 0x000000003FFF0030
BIOS32 Service Directory present.
Revision: 0
Calling Interface Address: 0x000FDA00
PCI Interrupt Routing 1.0 present.
Router ID: 00:01.0
Exclusive IRQs: None
Compatible Router: 8086:7000
Slot Entry 1: ID 00:01, on-board
Slot Entry 2: ID 00:02, slot number 1
Slot Entry 3: ID 00:03, slot number 2
Slot Entry 4: ID 00:04, slot number 3
Slot Entry 5: ID 00:05, slot number 4
Slot Entry 6: ID 00:06, slot number 5
Slot Entry 7: ID 00:07, slot number 6
Slot Entry 8: ID 00:08, slot number 7
Slot Entry 9: ID 00:09, slot number 8
Slot Entry 10: ID 00:0a, slot number 9
Slot Entry 11: ID 00:0b, slot number 10
Slot Entry 12: ID 00:0c, slot number 11
Slot Entry 13: ID 00:0d, slot number 12
Slot Entry 14: ID 00:0e, slot number 13
Slot Entry 15: ID 00:0f, slot number 14
Slot Entry 16: ID 00:10, slot number 15
Slot Entry 17: ID 00:11, slot number 16
Slot Entry 18: ID 00:12, slot number 17
Slot Entry 19: ID 00:13, slot number 18
Slot Entry 20: ID 00:14, slot number 19
Slot Entry 21: ID 00:15, slot number 20
Slot Entry 22: ID 00:16, slot number 21
Slot Entry 23: ID 00:17, slot number 22
Slot Entry 24: ID 00:18, slot number 23
Slot Entry 25: ID 00:19, slot number 24
Slot Entry 26: ID 00:1a, slot number 25
Slot Entry 27: ID 00:1b, slot number 26
Slot Entry 28: ID 00:1c, slot number 27
Slot Entry 29: ID 00:1d, slot number 28
Slot Entry 30: ID 00:1e, slot number 29
SMBIOS 2.5 present.
Structure Table Length: 450 bytes
Structure Table Address: 0x000E1000
Number Of Structures: 10
Maximum Structure Size: 255 bytes
```

biosdecode [opciones]	
OPTION	DESCRIPTION
-d	Read the memory device file /dev/mem
-V	Visualize the version.

> **dmidecode –t TYPE**
> TYPE
> bios
> system
> baseboard
> chassis
> processor
> memory
> cache
> connector
> slot

c)   Display the contents of the memory device.
>   biosdecode -t

STEP 6: Display the DMI devices.

It shows information about: Brand version and date of the BIOS, type of hardware supported, make and model of the board, type of socket, maximum memory size per slot and maximum supported. Information about ports and PCI slot, as well as those used and free.
>   Dmidecode

a)   Help.

        dmidecode --help

        dmidecode -h

b)   Default display.

        dmidecode

```
SMBIOS 2.5 present.
10 structures occupying 450 bytes.
Table at 0x000E1000.
Handle 0x0000, DMI type 0, 20 bytes
BIOS Information
Vendor: innotek GmbH
Version: VirtualBox
Release Date: 12/01/2006
Address: 0xE0000
Runtime Size: 128 kB
ROM Size: 128 kB
Characteristics:
ISA is supported
PCI is supported
Boot from CD is supported
Selectable boot is supported
8042 keyboard services are supported (int 9h)
CGA/mono video services are supported (int 10h)
ACPI is supported
Handle 0x0001, DMI type 1, 27 bytes
System Information
Manufacturer: innotek GmbH
Product Name: VirtualBox
Version: 1.2
Serial Number: 0
UUID: 345400D4-0150-4D45-BD0C-40E3FB67126D
Wake-up Type: Power Switch
SKU Number: Not Specified
Family: Virtual Machine
Handle 0x0008, DMI type 2, 15 bytes
Base Board Information
Manufacturer: Oracle Corporation
Product Name: VirtualBox
Version: 1.2
Serial Number: 0
Asset Tag: Not Specified
Features:
Board is a hosting board
Location In Chassis: Not Specified
Chassis Handle: 0x0003
Type: Motherboard
Contained Object Handles: 0
Handle 0x0003, DMI type 3, 13 bytes
Chassis Information
Manufacturer: Oracle Corporation
Type: Other
Lock: Not Present
Version: Not Specified
Serial Number: Not Specified
Asset Tag: Not Specified
Boot-up State: Safe
Power Supply State: Safe
Thermal State: Safe
Security Status: None
Handle 0x0007, DMI type 126, 42 bytes
Inactive
Handle 0x0005, DMI type 126, 15 bytes
Inactive
Handle 0x0006, DMI type 126, 28 bytes
Inactive
Handle 0x0002, DMI type 11, 7 bytes
OEM Strings
String 1: vboxVer_4.3.28
String 2: vboxRev_100309
Handle 0x0008, DMI type 128, 8 bytes
OEM-specific Type
Header and Data:
80 08 08 00 F0 04 23 00
Handle 0xFEFF, DMI type 127, 4 bytes
End Of Table
```

c)   Display all valid DMI types.

        dmidecode -t

d)   Display the DMI types

        dmidecode -t bios

        dmidecode -t system

        dmidecode -t baseboard

        dmidecode -t chassis

        dmidecode -t processor

        dmidecode -t memory

        dmidecode -t cache

        dmidecode -t connector

        dmidecode -t slot

dmidecode

OPTION	DESCRIPTION
-d	Read the memory from the /dev/mem file
-q	Less verbose output.
-s STRING	Display only the valid DMI strings.
-t TYPE	Only show the entries of given type.
-u	Do not decode the HEX inputs ..
-V	Visualize the version and exit.

Valid chains dmidecode -s

bios-vendor
bios-version
bios-release-date
system-manufacturer
system-product-name
system-version
system-serial-number
system-uuid
baseboard-manufacturer
baseboard-product-name
baseboard-version
baseboard-serial-number
baseboard-asset-tag
chassis-manufacturer
chassis-type
chassis-version
chassis-serial-number
chassis-asset-tag
processor-family
processor-manufacturer
processor-version
processor-frequency

e) Display all valid DMI chains.
        dmidecode -s
f) Display the value of the given DMI string.
        dmidecode -s   bios-version
        dmidecode -s   system-product-name
        dmidecode -s   system-uuid
        dmidecode -s   baseboard-product-name
        dmidecode -s   chasis-type
        dmidecode -s   processor-manufacturer

STEP 7: List part of our hardware.

a) It shows all the information of our processor and in the case of being dual-core, it appears as if they were two.

```
# less /proc/cpuinfo
processor       : 0
vendor_id       : GenuineIntel
cpu family      : 6
model           : 58
model name      : Intel® Core™ i7-3610QM CPU @ 2.30GHz
stepping        : 9
microcode       : 0x19
cpu MHz         : 2300.000
cache size      : 6144 KB
fpu             : yes
fpu_exception   : yes
cpuid level     : 5
wp              : yes
flags           : fpu vme de pse tsc msr pae mce cx8 apic sep mtrr pge mca
cmov pat pse36 clflush mmx fxsr sse sse2 syscall nx rdtscp lm constant_tsc
rep_good nopl pni monitor ssse3 lahf_lm
bogomips        : 4590.05
clflush size    : 64
cache_alignment : 64
address sizes   : 36 bits physical, 48 bits virtual
power management:
```

b) Displays information from memory. We can see how much memory we have and how much is available.

```
# less /proc/meminfo
Last login: Fri Aug 14 07:21:44 2015 from i7-dell.home
Linux 3.10.17.
When arguments fail, use a blackjack.
Ed "Spike" O'Donnell

root@192:~# less /proc/meminfo
MemTotal:        1012484 kB
MemFree:          499520 kB
Buffers:          208376 kB
Cached:           176188 kB
SwapCached:            0 kB
Active:           154600 kB
Inactive:         268924 kB
Active(anon):      38972 kB
Inactive(anon):     1040 kB
Active(file):     115628 kB
Inactive(file):   267884 kB
Unevictable:           0 kB
Mlocked:               0 kB
SwapTotal:       1047528 kB
SwapFree:        1047528 kB
Dirty:               200 kB
Writeback:             0 kB
AnonPages:         38944 kB
Mapped:            13624 kB
Shmem:              1044 kB
Slab:              67164 kB
SReclaimable:      58816 kB
SUnreclaim:         8348 kB
KernelStack:        1952 kB
PageTables:         4044 kB
NFS_Unstable:          0 kB
Bounce:                0 kB
WritebackTmp:          0 kB
CommitLimit:     1553768 kB
Committed_AS:     770180 kBxa
VmallocTotal:   34359738367 kB
VmallocUsed:       10024 kB
VmallocChunk:   34359723016 kB
AnonHugePages:     12288 kB
DirectMap4k:       12224 kB
DirectMap2M:     1036288 kB
```

STEP 8: Show the devices.

Show the devices.
        lshal
a) Help.
        lshal --help

b) Default display.

> lshal

c) Filtering of the Visualization. Pipes are used, because otherwise they give a very extensive output so we are going to limit them using a selective search and cut through the second column and display the output in alphabetical order.

> lshal | grep info.product | cut -d= -f2 | sort

STEP 9: Show the PCI devices.

> lspci

a) Help.

> lspci --help

b) Default display.

> lspci

```
00:00.0 Host bridge: Intel Corporation 440FX - 82441FX PMC [Natoma] (rev 02)
00:01.0 ISA bridge: Intel Corporation 82371SB PIIX3 ISA [Natoma/Triton II]
00:01.1 IDE interface: Intel Corporation 82371AB/EB/MB PIIX4 IDE (rev 01)
```

c) Basic display modes.

> lspci -m

```
00:00.0 "Host bridge" "Intel Corporation" "440FX - 82441FX PMC [Natoma]" -r02 ""
""
00:01.0 "ISA bridge" "Intel Corporation" "82371SB PIIX3 ISA [Natoma/Triton II]"
"" ""
00:01.1 "IDE interface" "Intel Corporation" "82371AB/EB/MB PIIX4 IDE" -r01 -p8a
"" ""
```

> lspci -mm

```
00:00.0 "Host bridge" "Intel Corporation" "440FX - 82441FX PMC [Natoma]" -r02 ""
""
00:01.0 "ISA bridge" "Intel Corporation" "82371SB PIIX3 ISA [Natoma/Triton II]"
"" ""
00:01.1 "IDE interface" "Intel Corporation" "82371AB/EB/MB PIIX4 IDE" -r01 -p8a
"" ""
```

> lspci -t

```
-[0000:00]-+-00.0
           +-01.0
           +-01.1
           +-02.0
           +-03.0
           +-04.0
           +-05.0
           +-06.0
           +-07.0
           +-08.0
           +-0b.0
           \-0d.0
```

lspci	
Basic viewing modes..	
-mm	Produces machine-rea dable output (-m only for an obsolete format).
-t	Show BUS tree.
Show options	
-v	It details (-vv for very detailed) drivers, to show kernel -k what drives each device.
-x	The standard part of the configuration space.
-xxx	Show hexadecimal the dump of the entire configuration space (dangerous; root).
-xxxx	Show hexadecimal space dump 4096 bytes extended configuration (root only) with -b the central view of the BUS (addresses and IRQ as seen by the BUS).
-D	Always show domain numbers.
The resolution of identification of devices to names.	
-n	Show numerical IDs
-nn	Show both the name and the numerical identification (names and numbers).
-q	Check the PCI identification database for unknown identifications through DNS.
-qq	Like the previous one, but re-query the entries cache locally.
-Q	Consult the PCI ID database for all IDs through DNS.

d) Display the contents of the entire configuration space in hexadecimal.

> lspci -x

```
00:00.0 Host bridge: Intel Corporation 440FX - 82441FX PMC [Natoma] (rev 02)
00: 86 80 37 12 00 00 00 00 02 00 00 06 00 00 00 00
10: 00 00 00 00 00 00 00 00 00 00 00 00 00 00 00 00
20: 00 00 00 00 00 00 00 00 00 00 00 00 00 00 00 00
30: 00 00 00 00 00 00 00 00 00 00 00 00 00 00 00 00
```

e) Displays the content of the entire configuration space in hexadecimal. (same)

> lspci -xx

f) Displays in hexadecimal the content of the entire extended space of 4096 bytes.

> lspci -xxx

```
00:00.0 Host bridge: Intel Corporation 440FX - 82441FX PMC [Natoma] (rev 02)
00: 86 80 37 12 00 00 00 00 02 00 00 06 00 00 00 00
10: 00 00 00 00 00 00 00 00 00 00 00 00 00 00 00 00
20: 00 00 00 00 00 00 00 00 00 00 00 00 00 00 00 00
30: 00 00 00 00 00 00 00 00 00 00 00 00 00 00 00 00
40: 00 00 00 00 00 00 00 00 00 00 00 00 00 00 00 00
50: 00 00 00 00 00 00 00 00 00 00 00 00 00 00 00 00
60: 00 00 00 00 00 00 00 00 00 00 00 00 00 00 00 00
70: 00 00 00 00 00 00 00 00 00 00 00 00 00 00 00 00
80: 00 00 00 00 00 00 00 00 00 00 00 00 00 00 00 00
90: 00 00 00 00 00 00 00 00 00 00 00 00 00 00 00 00
a0: 00 00 00 00 00 00 00 00 00 00 00 00 00 00 00 00
b0: 00 00 00 00 00 00 00 00 00 00 00 00 00 00 00 00
c0: 00 00 00 00 00 00 00 00 00 00 00 00 00 00 00 00
d0: 00 00 00 00 00 00 00 00 00 00 00 00 00 00 00 00
e0: 00 00 00 00 00 00 00 00 00 00 00 00 00 00 00 00
f0: 00 00 00 00 00 00 00 00 00 00 00 00 00 00 00 00
```

g) Display the identifier of the domain number.

> lspci -D

```
0000:00:00.0 Host bridge: Intel Corporation 440FX - 82441FX PMC [Natoma] (rev 02)
0000:00:01.0 ISA bridge: Intel Corporation 82371SB PIIX3 ISA [Natoma/Triton II]
```

```
0000:00:01.1 IDE interface: Intel Corporation 82371AB/EB/MB PIIX4 IDE (rev 01)
0000:00:02.0 VGA compatible controller: InnoTek Systemberatung GmbH VirtualBox Graphics Adapter
0000:00:03.0 Ethernet controller: Intel Corporation 82540EM Gigabit Ethernet Controller (rev 02)
0000:00:04.0 System peripheral: InnoTek Systemberatung GmbH VirtualBox Guest Service
0000:00:05.0 Multimedia audio controller: Intel Corporation 82801AA AC'97 Audio Controller (rev 01)
0000:00:06.0 USB controller: Apple Inc. KeyLargo/Intrepid USB
0000:00:07.0 Bridge: Intel Corporation 82371AB/EB/MB PIIX4 ACPI (rev 08)
0000:00:08.0 Ethernet controller: Intel Corporation 82540EM Gigabit Ethernet Controller (rev 02)
0000:00:0b.0 USB controller: Intel Corporation 82801FB/FBM/FR/FW/FRW (ICH6 Family) USB2 EHCI Controller
0000:00:0d.0 SATA controller: Intel Corporation 82801HM/HEM (ICH8M/ICH8M-E) SATA Controller [AHCI mode] (rev 02)
```

h) Display numerical identifiers.

```
lspci -n
00:00.0 0600: 8086:1237 (rev 02)
00:01.0 0601: 8086:7000
00:01.1 0101: 8086:7111 (rev 01)
00:02.0 0300: 80ee:beef
00:03.0 0200: 8086:100e (rev 02)
00:04.0 0880: 80ee:cafe
00:05.0 0401: 8086:2415 (rev 01)
00:06.0 0c03: 106b:003f
00:07.0 0680: 8086:7113 (rev 08)
00:08.0 0200: 8086:100e (rev 02)
00:0b.0 0c03: 8086:265c
00:0d.0 0106: 8086:2829 (rev 02)
```

i) Show the name and the numerical identification.

```
lspci -nn
00:00.0 Host bridge [0600]: Intel Corporation 440FX - 82441FX PMC [Natoma] [8086:1237] (rev 02)
00:01.0 ISA bridge [0601]: Intel Corporation 82371SB PIIX3 ISA [Natoma/Triton II] [8086:7000]
00:01.1 IDE interface [0101]: Intel Corporation 82371AB/EB/MB PIIX4 IDE [8086:7111] (rev 01)
00:02.0 VGA compatible controller [0300]: InnoTek Systemberatung GmbH VirtualBox Graphics Adapter [80ee:beef]
```

j) Display the result of the query the PCI identification database.

```
lspci -q
00:00.0 Host bridge: Intel Corporation 440FX - 82441FX PMC [Natoma] (rev 02)
00:01.0 ISA bridge: Intel Corporation 82371SB PIIX3 ISA [Natoma/Triton II]
00:01.1 IDE interface: Intel Corporation 82371AB/EB/MB PIIX4 IDE (rev 01)
00:02.0 VGA compatible controller: InnoTek Systemberatung GmbH VirtualBox Graphics Adapter
```

a) Displays the result of a new query in the cache of the entries locally. (In principle, like lspci -q).

```
lspci -qq
```

b) Visualize consulting the PCI ID database for all identifications through DNS.

```
lspci -Q
```

PRACTICE 13: Show files that exist in a Linux structure.

DESCRIPTION:

Standard order is ls, with this order we use search patterns, metacharacters.

Metacharacters.

Metacharacters	Meaning of Expression
*	Substitute between 0 or more characters.
?	Substitute between 0 and 1 character.
[]	Patterns or conditions.
[-]	Character range
[^] o [!]	Except that set of characters.
[a-z]	Character range between a and z both inclusive.
[1-6]	Numerical range between 1 and 6 both inclusive.
[1,3,5,6]	Set of elements (numerical).
[a,e,i,o,u]	Set of elements (vowels).
[^e*]	That does not contain the letter e.
[$a]	It should end in letter a.
{}	Replace string or strings inside the keys.
{mi,asa}	Files or directories that contain my or handle.
		Allows an alternative to choose between two expressions.
//	Delimit a regular expression.
\	Protect the next metacharacter.
{n}	Repetition of n times the previous character or subexpression.
{n,}	Repetition of n times the previous character or subexpression.
{n,m}	Repeat between n and m times the previous character or subexpression.

METACHARACTERS NOT PRINTS

METACHA-RACTERS	MEANING
\a	Beep, the BEL character (07 in hexadecimal).
\e	Escape (1B in hexadecimal).
\cx	"control-x", where x is the corresponding character.
\f	New page (0C hexadecimal).
\n	New line (0A hexadecimal).
\r	Return carriage (0D hexadecimal).
\t	Tabulator (09 hexadecimal).
\xhh	Character with hh hexadecimal code.
\ddd	Character with ddd code in octal.

Clases

You can specify the character classes according to several syntaxes, POSIX, traditional or Unicode.

According to the POSIX class syntax, [: class:] is indicated

Classe	Meaning
[:alpha:]	Alphabetical character
[:alnum:]	Alphanumeric character
[:ascii:]	Character ASCII
[:blank:]	Space, includes tabulator (also \s according to the traditional syntax).
[:cntrl:]	Control character.
[:digit:]	Graphic character
[:graph:]	Lowercase letter.
[:lower:]	One digit (also \d according to the traditional syntax).
[:print:]	Printable character.
[:punct:]	Punctuation character.
[:space:]	Space (also \s according to the traditional syntax).
[:upper:]	Capital letter.
[:word:]	Word (also \w according to the traditional syntax).
[:xdigit:]	Hexadecimal digit.

OTHER METACARACTERS

Metacarácter	Meaning
\D	Any character other than a decimal digit (equivalent to [^: digit:])
\S	Any character that is not a blank space (equivalent to [^: blank:]
\w	Any character of a word.
\W	Any character that is not a "word".

Examples:

Expression	Meaning
/[a-z]/	A lowercase letter. The "-" indicates a range, which in this case starts at "a" and ends at "z".
/[A-Z]/	An uppercase letter.
/[0-9]/	One digit
/[,'¿!¡;:.?]/	A punctuation character.
/[A-Za-z]/	A letter except accented and ñ.
/[A-Za-z0-9]/	A letter, except accented and ñ, or a digit
/[^a-z]/	Any character except one lower case letter.
/[^0-9]/	Any character except a number.

LITERALES

Consists of: two literals: tn: and (EN) mydomain.dom (the "." Is protected). An anchor, the $ end of line indicating two sets of characters [] that represents a blank space and [^ (EN)] that indicates everything that is not (EN). The quantifiers * that indicate in this case possible blank spaces and + that indicates repetition of one or more times the previous character, in this case everything that is not (EN). The metacharacters * and + are discussed in more detail in their specific section.

```
/En:[ ]*[^(EN)]+(EN)midominio\.dom$/
```

Metacharacter as literal

Using a metacharacter as a literal will have to protect it with a backslash ("\"). A particular case is to use regular expressions within a shell, for example a shell script. In this case we should perform in some cases a double protection with the contrabarra, one to protect it in shell and another to protect it in the regular expression.

   x="The variable \\$X is defined"

It is stored, "The variable $ X is defined". This string as a regular expression does not have the $ protected sign.

If we wanted to store "The variable \ $ X is defined" you have to put

   x=" The variable \\\\$x is defined"

STEP 1: Command ls.

View or list the information content of the current directory.

   ls

a) Help.

   ls --help

b) Default display.

   ls

c) Visualize in large or long format.

   ls –l

d) Display all files and directories (include hidden ones).

   ls -a

   ls -la

e) Display the current and active directory.

   ls -d

   ls -dl

   pwd

Current directory ./

Executable files, outside of the search path you have to precede it from ./.

   ls -li /etc

f) List by modification time of an inode

   ls -t /sbin

   ls -tl /bin

g) List the contents of the directory according to the version (update files).

   ls -v /bin

   ls -lv /bin

h) Display all hidden files or directories (without ignoring "." "..").

   ls -la /bin

i) Display all hidden files or directories, including less (.) (..)

   ls -lA /bin

  Do not list the backup files (~).

   cd /mnt/local

   ls -B

   ls -lB

j) Reverse the display order.

   ls -lB

   ls -lB -r

k) Display the file by Size.

   ls -lS

l) Invest, the opposite of the size greater than the smallest.

   ls -lS -r

m) Display a column with the block size.

   ls -ls

n) The number of blocks is the first column that appears

   ls -lsS

   ls -lsSr

o) Reverse the selection.

   ls -lsSi

  A column (1st) is added with the inode inode number.

   ls -lsSir

p) List the files by creation date or last modification.

   ls -t

   ls -lt

q) Deactivate the color and display in columns.

   ls -f

   ls -lf

r) Display the files in long format, omitting the owner's column.

   ls -g

   ls -lg

ls

Syntax: ls [-aAcCdfFgilLmnpqrRstux1] [path ...]

OPTION	DESCRIPTION
-a	List all entries.
-F	It puts '/' at the end of directories, '*' at the end of executables and '@' at symbolic links, '|' FiFo.
-i	Show the inode number in column 1
-l	Long list.
-n	With -l shows UID / GID.
-p	Put '/' at the end of the directories.
-r	Invests the sense of order.
-s	Shows size in blocks.
-u	Sorting sense by the last access date.
-1	Force the format of a file name on each line.
-A i	Same as -a but it does not list the directories '.' And '...'
-L	Follow the symbolic links.
-o	Like -l but it does not show group.
-q	Displays the non-displayable characters.
-R	Recursively listing directories.
-t	Sort by dates
-x	The output is made in horizontally ordered multicolumn.

It visualizes in columns and establishes the difference by colors and its termination.

 / Directories (dark blue).

 * Executable (green).

 ~ Backup copies (.bak, bk!).

 | Pipes.

 = socket (purple).

 -> Link (light blue).

s) Display in long format the files omitting the group column.

    ls -G
    ls -LG

t) Visualize in multicolumn ordered horizontally.

    ls -x

u) Visualize in vertically ordered multicolumn. Default display.

    ls

v) Display in wide format equal -l, and does not show groups.

    ls -o

w) Display the contents of all the directories recursively, and in a broad format.

    ls -lR

x) Display by the date of the last access, in ample format.

```
ls -lu
total 2260
drwxr-xr-x    5 root root      4096 Aug  7 04:40 ConsoleKit/
rw-r—r--      1 root root      4593 Aug  9 14:49 DIR_COLORS
rw-r—r--      1 root root        26 Aug  9 11:42 HOSTNAME
```

y) Display the UID / GUID.

```
ls —ln
total 1124
rw-r—r--.     1  0  0        16 ago  6 01:54 adjtime
rw-r—r--.     1  0  0      1518 jun  7 2013 aliases
rw-r—r--.     1  0  0     12288 ago  6 03:22 aliases.db
drwxr-xr-x.   2  0  0      4096 ago  6 01:35 alternatives
rw-------.    1  0  0       541 jul 30 2014 anacrontab
```

z) Display in large format and add one of these symbols to the end of the file name (* / => @ |).

```
ls -lF
drwxr-xr-x    2 root root      4096 Apr 18 2013 auto.master.d/
rw-r—r--      1 root root       524 Apr 18 2013 auto.misc
rwxr-xr-x     1 root root      1260 Apr 18 2013 auto.net*
rwxr-xr-x     1 root root       687 Apr 18 2013 auto.smb*
```

aa) Display in large format and the i-node number appears

```
ls -li
 33554589 -rw-r—r--.  1 root root        23 abr  1 00:27 system-release-cpe
 33907631 -rw-------.  1 tss  tss       6411 jun 10 2014 tcsd.conf
 33555576 drwxr-xr-x.  2 root root         6 jun 11 2014 terminfo
100951830 drwxr-xr-x.  2 root root         6 mar  6 06:48 tmpfiles.d
101208979 drwxr-xr-x.  2 root root        67 ago  6 01:35 tuned
```

bb) Display add / to the end of the directory name.

```
ls -lp
drwxr-xr-x.   2 root root        32 ago  6 01:35 wpa_supplicant/
drwxr-xr-x.   5 root root        54 ago  6 01:34 X11/
drwxr-xr-x.   4 root root        36 ago  6 01:34 xdg/
drwxr-xr-x.   2 root root         6 jun 10 2014 xinetd.d/
drwxr-xr-x.   6 root root      4096 ago  6 01:35 yum/
rw-r—r--.     1 root root       970 mar  9 21:39 yum.conf
drwxr-xr-x.   2 root root      4096 mar  9 21:39 yum.repos.d/
```

cc) Display all the files that end in a character.

```
[root@localhost bin]# ls  -l  [$o]*
ls: no se puede acceder a []*: No existe el fichero o el directorio
[root@localhost bin]# ls  -l  [$a]*
ls: no se puede acceder a []*: No existe el fichero o el directorio
[root@localhost bin]# ls  -l  | grep  'ep'
```

dd) Display all the files that have in the first character an a, e, or any other character included between letter b and p.

```
ls -l /bin/[a,e,b-p]*
[root@localhost bin]# ls  -l  [a,e,b-p]*
rwxr-xr-x.    1 root root     29016 mar  5 23:27 addr2line
rwxr-xr-x.    1 root root        29 mar  5 23:06 alias
lrwxrwxrwx.   1 root root         6 ago  6 01:35 apropos -> whatis
rwxr-xr-x.    1 root root     58472 mar  5 23:27 ar
rwxr-xr-x.    1 root root     33048 jun 10 2014 arch
```

ee) Display any file that you have in the first character from the letter a to the lowercase m or from the letter A to the M and any other character.

```
ls -l /bin/[a-m,A-M]*
[root@localhost bin]# ls  -l  [a-m,A-M]*
rwxr-xr-x.    1 root root     29016 mar  5 23:27 addr2line
rwxr-xr-x.    1 root root        29 mar  5 23:06 alias
lrwxrwxrwx.   1 root root         6 ago  6 01:35 apropos -> whatis
rwxr-xr-x.    1 root root     58472 mar  5 23:27 ar
rwxr-xr-x.    1 root root     33048 jun 10 2014 arch
rwxr-xr-x.    1 root root    365200 mar  5 23:27 as
```

ff) Display all files and directories whose first character is an e or a.

```
ls -l /bin/[$a,e]*
[root@localhost bin]# ls  -l  [$a,e]*
rwxr-xr-x.    1 root root       320 jun 10 2014 easy_install
rwxr-xr-x.    1 root root       328 jun 10 2014 easy_install-2.7
```

```
rwxr-xr-x.   1 root root  33040 jun 10  2014 echo
rwxr-xr-x.   1 root root    158 mar  6 01:17 egrep
rwxr-xr-x.   1 root root  45640 mar  6 06:59 eject
rwxr-xr-x.   1 root root  32920 mar  5 23:27 elfedit
rwxr-xr-x.   1 root root  28960 jun 10  2014 env
rwxr-xr-x.   1 root root  36816 jun 10  2014 envsubst
rwxr-xr-x.   1 root root 147880 jun  9  2014 eqn
lrwxrwxrwx.  1 root root      2 ago  6 01:34 ex -> vi
rwxr-xr-x.   1 root root  33216 jun 10  2014 expand
rwxr-xr-x.   1 root root  37384 jun 10  2014 expr
```

gg) Display all files and directories whose first character is an e or a.

    ls -l /bin/[$^a*]

    ls -l /bin/[$^a*]*

    ls -l /bin/[$^a]*

```
rwxr-xr-x.   1 root root  29016 mar  5 23:27 addr2line
rwxr-xr-x.   1 root root     29 mar  5 23:06 alias
lrwxrwxrwx.  1 root root      6 ago  6 01:35 apropos -> whatis
rwxr-xr-x.   1 root root  58472 mar  5 23:27 ar
rwxr-xr-x.   1 root root  33048 jun 10  2014 arch
```

hh) Visualize that they begin with or.

    ls -l /bin/[$^o]*

```
rwxr-xr-x.   1 root root 224280 mar  5 23:27 objcopy
rwxr-xr-x.   1 root root 332248 mar  5 23:27 objdump
rwxr-xr-x.   1 root root  66320 jun 10  2014 od
rwxr-xr-x.   1 root root 190816 jun 10  2014 oldfind
lrwxrwxrwx.  1 root root      6 ago  6 01:35 open -> openvt
rwxr-xr-x.   1 root root 508680 mar  6 05:49 openssl
rwxr-xr-x.   1 root root  19928 mar  6 04:13 openvt
rwxr-xr-x.   1 root root   5618 jun 10  2014 os-prober
```

ii) Display all files and directories containing a string or part, containing (ae), (la), (sy) and any other character.

    ls -l /bin/{ae,la,sy}*

```
ls: you can not access ae *: There is no file or directory
rwxr-xr-x.   1 root root  19568 jun 10  2014 last
lrwxrwxrwx.  1 root root      4 ago  6 01:34 lastb -> last
rwxr-xr-x.   1 root root  15392 mar  6 06:32 lastlog
rwxr-xr-x.   1 root root  28952 jun 10  2014 sync
rwxr-xr-x.   1 root root 357752 mar  6 06:48 systemctl
```

jj) Visualize the files in a single column, in simple format.

    ls -1

```
[
addr2line
alias
apropos
```

kk) Display the files in a single large format column displaying the UID/GUID.

    ls -1 -n

PRACTICE 14: Processing of files in Linux.

DESCRIPCIÓN:

Unix systems, and Linux in particular, have advanced tools that allow the manipulation of text files to extract information and modify them. This is really important since most of the configuration files of a Linux system are text files that we usually have to manipulate.

Manipulate information of the files:

- Compare files.
- Compress / decompress files.
- Compact files.

STEP 1: Compare EQUAL files.

The binary comparison is made until the end of the files, as long as the number of bytes to be compared is the same.

```
cmp
```

a) Help.
```
man  cmp
cmp  --help
info  cmp
```

b) Compare files.
   Use addresses
```
ls -l /bin  > salida1
ls -l sbin  > salida2/
cmp  -c salida1  salida2
```

c) Display text files.
```
cat
more
less
cat  salida1
cat  salida2
```

c.1) Display text files.
```
cat eje010
tac  eje010
cat  salida1 > salida3
tac  salida1 > salida4
ls -l
cmp -c  salida1  salida3
cmp -c  salida1  salida4
```

d) Display differences at the byte level.
```
cmp  -b salida1  salida4
```

cmp [options..] file1 file2	
OPTION	DESCRIPTION
-c	Show the different octets as characters.
-l	Displays the number of octets (decimal) and the value of different octets (octal) for each difference.
-s	It does not show anything for different files, it returns the exit status only.

cmp --> first compare is the same size assumes that it contains the same information.

STEP 2: Show differences.

It allows to compare two files line by line and informs us of the differences between both files.
```
diff
```

a) Help.
```
diff --help
```

b) Show the differences between files by default.
```
diff  salida1 salida4
cat salida1 >salida5
cat salida1 >>salida5
```

c) Display page shape differences.
```
diff  -l salida1  salida5
```

d) It allows comparing directories (recursive).
```
diff  -r /bin  /sbin
```

e) Compare the files next to each other, ignoring the blank spaces.
```
diff  -by salida1  salida5
```

f) Compare the files next to each other, ignoring the blank spaces.
```
diff  -iy salida1 salida5
```

diff	
Sintaxis:	diff [options...] file1 file2
OPTION	DESCRIPTION
-a	Treat all files as text and compare them line-by-line.
-b	Ignore changes in the amount of white spaces.
-c	Use the context output format.
-e	Makes the output a valid ed script.
-H	Use heuristics to accelerate the handling of large files that have small, scattered changes.
-i	Ignore changes between uppercase and lowercase, consider them equivalent
-n	Show in RCS format, such as -f except that each command specifies the number of affected lines.
-q	Show diffs in RCS format, such as -f except that each command specifies the number of affected lines.
-r	When comparing directories, it compares repeatedly any subdirectory found.
-s	Reports when two files are the same.
-w	Ignore white space when comparing lines.
-y	Use the output format one next to the other.

STEP 3: Show differences comparing files by columns.
```
cd /mnt/local
cat > datos
```

a) Help.
```
comm --help
```

b) Default.
```
comm -1  -1 datos datos1
```

comm	
Sintaxis:	comm [options]... file1 file2
OPTION	DESCRIPTION
-1	Delete exclusive lines from left file
-2	Delete exclusive lines from the correct file
-3	Delete the lines that appear in both files.

c) Compare by columns.
    comm -1 -2 salida salida1
    comm -3 salida salida1

STEP 4: Sort text files.
        sort
a) Help.
        sort --help
b) Default.
        sort datos
        sort datos > salida
        cat salida
        sort datos1 >salida1

sort	
Sintaxis:	sort [options] name_of_file
OPTION	**DESCRIPTION**
-r	Order in reverse order.
-u	If the line is duplicated, it shows only once.
-o name_file	Send the ordered output to a file.

STEP 5: Compression / Decompression of files.
        gzip
a) Help.
        gzip --help
b) Default.
        ls -lR / > comprime
        ls -lR / > comprime &2>error
        ls -l
        gzip   comprime
        ls -l
        The origin -> compress disappears I am left with the file compressed with the extension .gz
c) Unzip directly with gzip.
        gzip -d comprime.gz
        ls -l
    The compressed file disappears and the uncompressed file appears, without the .gz characters.

gzip	
Syntax: gzip	
OPTION	**DESCRIPTION**
-1	Information about the compression of a .gz file
-S	Change the .gz extension
-r	Compress directories.
-c	Redirect the output of a compressed file.
-t	Check the integrity of a compressed file
-d	Unzip .gz files

d) Redirect the output of a compressed file. The gzip command does not carry files, then it reads from the standard input and to save the compression we have to send it to the standard output and redirect it to a file.
        man ln | gzip -c > salida.ln.man.gz
e) Check the integrity of a compressed file.
        gzip -t comprime.gz
f) Unzip order.
        gzip   comprime
        ls -l
        gunzip   comprime.gz
        ls -l

tar		
Syntax: tar <options> file_a_create <files_to_adjust>		
OPTION	**DESCRIPTION**	
-c	Tell tar to create a file.	
-v	Tell tar to show what it is packing.	
-f	Tell tar that the next argument is the name of the file.tar	
-x	Tell tar to unzip the file.tar.	
-v	Tell tar to show what you are unpacking.	
-t	List the contents of the .tar file	

g) Compress a directory, including the contents of a subdirectory ..
        gzip -r   tmp
h) Change the output extension of the files.
        gzip -S .gzip ls.man rm.man

STEP 6: Compress files and directories (Pack up | Unpack).
        tar
a) Help.
        tar --help
b) Compress using the .tar format (if streamer backup tapes are used, compression is done with bar).
        tar cf salida01.tar . --> mal
        tar cf salida01.tar /etc
        ls -l
c) Compress a compacted file.
        ls -l
        gzip salida01.tar
        ls -l
        gunzip salida01.tar.gz
d) Visualize the contents of a compacted file.
        tar tf salida01.tar
        Descomprimir--> Descompactar
        tar xf salida01.tar   ./estudiar
e) Add information to a compacted file.
        tar rf salida01.tar /bin
        tar -tf salida01.tar
        tar rf salida01.tar ./estudiar

```
        tar  tf  salida01.tar
        rm -r  estudiar
        tar  xf  salida01.tar
```
f)   Update a compacted file.
```
        tar  uf  salida01.tar ./nuevo
```
g)   Compress in .gz format.
```
        tar  czvf  salida02.gz  /etc
        mv salida02.tar salida02.gz
```
h)   Delete directories.
```
        rm -rf  etc
        ls   -l
        tar  xzvf  salida02
```
i)   Unzip a .tar .gz file.
```
        tar  xvf  copia.tar  ./local
        tar  xzvf  copia2.gz  ./local1
```
Example:
```
        tar  czvf  salida001.gz  /bin
        tar  czvf  salida002.gz  /etc
        ls  -l /    > texto01
        ls  -l /etc > texto02
        gzip  texto01
        gzip  texto02
```
j)   Help.
```
        bzip2  --help
```
i)   To compress files in bz2 format, the following command is used:
```
        bzip text02
```
j)   To decompress .bz2 files, the following command is used:
```
        bzip2 -d fichero.bz2
```

STEP 7: Manipulate compressed text files.
```
        zcat, zmore, zcmp, zdiff
```
a)   Help.
```
        zcat      --help
        zmore     --help
        zcmp      --help
        zdiff     --help
```
b)   Display the contents of a compressed text file.
```
        zcat   texto01.gz
        zcat   texto02.gz
        zcat   texto01.gz | less
        zcat   texto02.gz | less
```
c)   Display the contents of a compressed file, by screen, by scroll.
```
        zless texto01.gz
        zless texto02.gz
```
d)   Display the contents of a compressed text file slowly (screen to screen) scroll.
```
        zmore  texto01.gz
        zmore  texto02.gz
```
e)   Compare the same compressed files.
```
        zcmp texto01.gz  texto02.gz
```
f)   Compare different text files.
```
        zdiff  texto01.gz  texto02.gz
```
Example:
```
        cat  /etc/passwd  >salida03
        cat  /etc/group   >salida04
        gzip  salida03
        gzip  salida04
        zcat   salida03.gz
        zmore salida03.gz
        zcat   salida04.gz
        zmore salida04.gz
        zcmp  salida04.gz  salida03.gz
        zdiff  salida03.gz  salida04.gz
```
Shows the line number
```
        zdiff -n salida03.gz  salida04.gz
```
Show and mark the differences.
```
        zdiff -c salida03.gz  salida04.gz
```

bzip2
Syntax: bzip2 [options] file

OPTION	DESCRIPTION
-d	Force to decompression.
-z	Force to compression.
-k	Keep (not delete) the input files.
-f	Overwrite the existing output files.
-t	Test the integrity of the compression file.
-c	Standard output out..
-q	Remove non-critical error messages.
-v	Visualize the compression.
-s	Use less memory (maximum 2500K).
-1 .. -9	Set the block size between 100k ... 900k.
--fast	Alias for -1.
--best	Alias for -9

ORDER	USE
tar	Pack: tar -cvf file.tar / pract / my / all / Unpack: tar -xvf file.tar View content tar -tf file.tar
gz	Compress: gzip -9 file Unzip: gzip -d file.gz
bz2	Compress: bzip file Unzip: bzip2 -d file.bz2 gzip or bzip2 only compress files [not directories, for that there is tar]. To compress and file at the same time you have to combine the tar and the gzip or the bzip2 as follows, next steps of the table
tar.gz	Compress files: tar -czfv file.tar.gz files Unzip: tar -xzvf file.tar.gz See content: tar -tzf file.tar.gz
tar.bz2	Compress: tar -c files | bzip2> filear.bz2 Unzip: bzip2 -dc file.tar.bz2 | tar -xv See content: bzip2 -dc file.tar.bz2 | tar -t
zip	Compress: zip file.zip files Unzip: unzip file.zip View content: unzip -v file.zip
lha	Compress: lha -a file.lha files Unzip: lha -x file.lha See content: lha -v file.lha See content: lha -l file.lha
arj	Compress: arj to file.arj files Unzip: unarj file.arj Unzip: arj -x file.arj View content: arj -v file.arj View content: arj -l file.arj
zoo	Compress: zoo to file.zoo files Unzip: zoo -x file.zoo See content: zoo -L archivo.zoo View content: zoo -v file.zoo
rar	Compress: rar -a file.rar files Unzip: rar -x file.rar View content: rar -l file.rar View content: rar -v file.rar

PRACTICE 15: Create soft and hard links or links in Linux.
DESCRIPTION:
What are they? What are they for?

Linux, every file in the system is represented by an inode. An inode is nothing more than a block that stores information from the files, so that each inode can be associated with a name. At first glance it would seem that we can not associate several names with the same file, but thanks to the links this is possible.

Symbolic Links

A symbolic link (soft link, or shortcut) is a special file that contains a path name. Thus, soft links can point to files in different file systems (possibly mounted by NFS from different machines, removable drives), and do not have to point to files that actually exist.

A symbolic link allows to give a file the name of another, but does not link the file with an inode, that is, in fact what we do is link directly to the file name. Symbolic links are widely used for shared libraries.

To understand it better, a "symbolic link" is just a reference (link) to a folder (directory) or file that is located in a different physical location.

Hard links

The hard links what they do is associate two or more files sharing the same inode. This makes each hard link an exact copy of the rest of the associated files, both data and permissions, owner, etc. This also implies that when changes are made in one of the links or in the file, this will also be done in the rest of the links.

In GNU / Linux systems, hard links have several limitations:
- Hard links can only be made to files, and not to directories.
- They can not be expanded through different file systems. This means that you can not create a permanent link from /usr/bin/bash to /bin/bash if your / and /usr directories belong to different file systems.

Conclusion:
- Symbolic links can be made with files and directories, hard links with files only.
- Symbolic links can be made between different file systems, hard links can not.
- In the symbolic links if the original file or directory is deleted the information is lost, in the hard links there is no.
- Hard links are copies of the originals that share the inode number, while the symbolic links are mere pointers.
- There are different ways to delete links; unlink, rm7.

Orders:
ln
symlinks
unlink

STEP 1: Create direct access to files and directories.

Create hard links and symbolic links to files
ln

a) Help.
    ln --help
    man ln
    info ln
b) Default assignment.
    cd /mnt/local
    ls -l
    ln eje011 eje012
c) By default a hard link is created, a copy of the original file.
    ls -l
    Edit the file eje011.
    nano eje011
d) Observe the result on axis011 and axis012.cat eje011
    cat eje012
e) Create symbolic links (soft links).
    ln -s eje012 eje014
    ln -s eje014 eje015
    ln -s eje014 eje016
    ln -s eje012 eje017
    ls -l
    ln eje011 eje013
    rm eje011
    ls -l
    rm eje012
    cat eje014
f) Create symbolic access to a directory.
    ln -s deportes futbol
    ln eje013 eje012

ln	
Syntax:	**ln [options] existing_archive_name (or directory) new_file_name (or directory)**
OPTION	**DESCRIPTION**
-f	Link files without prompting the user, even if the file mode prohibits writing. This is by default if the standard input is not a terminal.
-n	It does not overwrite existing files.
-s	It is used to create soft, soft links.

The symbolic links, go from blue to red.

STEP 2: Search from a directory, symbolic links.

Check symbolic links in the file system.

symlinks

a) Help.

symlinks --help

man symlinks

info symlinks

b) Search for symbolic, relative links in a directory.

symlinks -v /bin

symlinks -v /mnt/local/deportes

symlinks -v .

apt-get install symlinks

c) Search for hard links and change through symbolic links.

symlinks -c /bin

d) Search recursively -r.

symlinks -r /

symlinks -r .

symlinks -r -v .

```
symlinks
    apt-gets  install  symlinks
    yum       install  symlinks
```

```
ln
symlinks [-cdorstv]  list-directory
```

OPTION	DESCRIPTION
-c	They change absolute / messy links to relative.
-d	Remove hanging links
-o	Inform about the links through the file systems.
-r	Recursive in subdirectories.
-s	Shorten long links (shown in the output only when -c not specified).
-t	Show what would be done by -c.
-v	Show all symbolic links.

e) Report the links through the file systems. symlinks -o / bin

```
absolute: /bin/iptables-xml -> /usr/sbin/xtables-multi
dangling: /bin/rpmquery -> ../../bin/rpm
dangling: /bin/rpmverify -> ../../bin/rpm
absolute: /bin/ld -> /etc/alternatives/ld
absolute: /bin/mailq -> /etc/alternatives/mta-mailq
absolute: /bin/rmail -> /etc/alternatives/mta-rmail
dangling: /bin/mailq.postfix -> ../../usr/sbin/sendmail.postfix
dangling: /bin/newaliases.postfix -> ../../usr/sbin/sendmail.postfix
absolute: /bin/newaliases -> /etc/alternatives/mta-newaliases
messy:    /bin/slogin -> ./ssh
```

f) Delete hanging links.

symlinks -d

g) Show what would be done by -c.

```
symlinks -t /bin
absolute: /bin/iptables-xml -> /usr/sbin/xtables-multi
changed:  /bin/iptables-xml -> ../usr/sbin/xtables-multi
dangling: /bin/rpmquery -> ../../bin/rpm
dangling: /bin/rpmverify -> ../../bin/rpm
absolute: /bin/ld -> /etc/alternatives/ld
changed:  /bin/ld -> ../etc/alternatives/ld
absolute: /bin/mailq -> /etc/alternatives/mta-mailq
changed:  /bin/mailq -> ../etc/alternatives/mta-mailq
absolute: /bin/rmail -> /etc/alternatives/mta-rmail
changed:  /bin/rmail -> ../etc/alternatives/mta-rmail
dangling: /bin/mailq.postfix -> ../../usr/sbin/sendmail.postfix
dangling: /bin/newaliases.postfix -> ../../usr/sbin/sendmail.postfix
absolute: /bin/newaliases -> /etc/alternatives/mta-newaliases
changed:  /bin/newaliases -> ../etc/alternatives/mta-newaliases
messy:    /bin/slogin -> ./ssh
changed:  /bin/slogin -> ssh
```

STEP 3: Deleting hard links.

unlink

a) Help.

unlink --help

b) As with symbolic links, we can use two commands to erase hard links:

unlink /home/baldo/enlace-duro2

c) Use rm to delete links.

rm /home/baldo/enlace-duro2

Analyze: root@alumno-svr:/home/alumno#

ls l > salida

ln salida salida1

ln salida1 salida2

ln -s salida2 salida3

ln -s salida3 salida4

ls -l

rw-r—r--	3 root root	102 oct 12 00:27	salida
rw-r—r--	3 root root	102 oct 12 00:27	salida1
rw-r—r--	3 root root	102 oct 12 00:27	salida2
lrwxrwxrwx	1 root root	7 oct 12 00:29	salida3 -> salida2
lrwxrwxrwx	1 root root	7 oct 12 00:29	salida4 -> salida3

The second column indicates the number of inodes that make up the hard link, the number 3 indicates that there are 3 hard links or three copies of the same inode and the same content, the fifth column indicates the size, you can see the size of the first three files is the same, the date and time is the same, the name is different, all three are updated if one of them is modified.

The last two indicate in the first column the first character l -> soft link, there is no inode replication, but a reference to the same iniodo, output 2 indicates that soft link is output3. It is referenced from two files to the same inode, if the original file is deleted the link to the inode disappears.

STEP 4: List attributes of a directory.

The lsattr command is used to list the attributes of specified directories or files.

       lsattr

a)   Help.

       lsattrv --help

b)   List the default attributes.

       lsattr

c)   List the attributes of directories in depth recursively.

       lsattr -R

d)   List all attributes including hidden (. ..)

       lsattr -a

e)   List directories as if they were files, instead of listing their contents.

       lsattr -d

f)   List directories as if they were files, instead of listing their contents.

       chattr +i test.txt

g)   Display the established attributes.

lsattr	
Syntax:	**lsattr [options]**
OPTION	**DESCRIPTION**
-R	List repeatedly the attributes of the directories and their contents.
-a	List all files in directories, including files that start with `.'.
-d	List directories as other files, instead of listing their contents

```
root@192:~# lsattr texto.txt
-------------e-texto.txt
root@192:~# chattr +i texto.txt
root@192:~# lsattr texto.txt
----i--------e-texto.txtç
root@192:~# lsattr -R
-------------e-- ./slapt-get
./slapt-get:
-------------e-- ./slapt-get/slapt-get-0.10.2r-x86_64-1.tgz
-------------e-- ./fich005
-------------e-- ./fich001
----i--------e-- ./texto.txt
-------------e-- ./fich002
-------------e-- ./fich003
-------------e-- ./fich004
root@192:~# lsattr -a
-------------e-- ./..
-------------e-- ./slapt-get
-------------e-- ./.xinitrc-backup
-------------e-- ./.xsession-backup
-------------e-- ./.
-------------e-- ./.gnupg
-------------e-- ./.xinitrc
-------------e-- ./fich005
-------------e-- ./.bash_history
-------------e-- ./.xsession
-------------e-- ./fich001
----i--------e-- ./texto.txt
-------------e-- ./fich002
-------------e-- ./fich003
-------------e-- ./fich004
root@192:~# lsattr -d
-------------e-- .
```

NOTE: chattr included in the e2fsprogs package, which is installed by default in all GNU / Linux distributions. It includes other tools such as e2fsck, e2label, fsck.ext2, fsck.ext3, fsck.ext4, mkfs.ext2, mkfs.ext3, mkfs.ext4, tune2fs and dumpe2fs, ...

STEP 5: Change attributes of a directory.

Change the attributes of the ext2, ext3 and ext4 file systems.

       chattr

a) Help.

       chattr --help

       chattr

b) Change the read-only file attribute.

       chattr +i fich001

       lsattr

c) Remove the read-only attribute.

       chattr -i   fich001

       chattr

d) Can not open the file to write.

       chattr +a fich001

e) Open the file for writing.

    chattr -a fich001

f) Changes to the file are written to the disk at the same time.

    chattr +S fich001

g) Do not activate for dump copies.

    chattr +d   fich001

h) File written with Journal.

    chattr +j fich001

i) Write the blocks used on the disk with zeros.

    chattr +s fich001

j) Permanently save.

    chattr +u fich001

k) Establish several attributes.

    chattr +a +i +S fich001

l) If the Ext3 file system, it is established that the file fich001 will only have the attributes **a**, **A**, **s** and **S**.

    chattr =aAsS fich001

---

chattr

Syntax: chattr [-RVf] [-+=AaCcDdeijSsTtu] [-v version] files...

OPERATORS	DESCRIPTION
+	It causes the specified attributes to be added to the existing attributes of a file.
-	Makes the specified attributes of the existing attributes of a file deleted.
=	Makes the existing attributes replaced by the specified attributes.

ATTRIBUTE	DESCRIPTION
i	Makes the file read-only.
A	It establishes that the date of the last access (atime) is not modified
a	Open the file for writing, not writing.
c	It establishes that the file is automatically compressed on the disk by the kernel of the operating system. When reading this file, the data is decompressed. Writing that file compresses the data before storing it on disk.
D	When it is a directory, it establishes that the data is written synchronously on the disk. That is, the data is written immediately instead of waiting for the corresponding operation of the operating system. It is equivalent to the mount dirsync option, but applied to a subset of files.
d	It establishes that the file is not a candidate for backup when using the dump tool.
e	Indicates that the file or directory uses extensions (extents) for the block mapping of the storage unit, particularly of Ext4 file systems. It is important to know that chattr is unable to eliminate this attribute.
i	It establishes that the file will be immutable. That is, the file is prevented from being deleted, renamed, symbolic links can be pointed to it or writing data to the file.
j	In the ext3 and ext4 filesystems, when mounted with the data = ordered or data = writeback options, it is established that the file will be written in the journal log (Journal). If the file system is mounted with the option data = journal (default option), the entire file system is written to the journal record and therefore the attribute has no effect.
s	When a file has this attribute, the blocks used in the hard disk are written with zeros, so that the data is not any. It is the surest way to delete data.
S	Changes to the file are written to the disk at the same time. It is equivalent to the sync option of mount.
u	When a file with this attribute is deleted, its contents are saved, allowing you to recover the file with tools for that purpose

OPTION	DESCRIPTION
-R	Changes the attributes of directories and their contents in a descending manner. The symbolic links that are found are ignored.
-V	More descriptive charttr output, also showing the program version.
-v	See the version number of the program.

PRACTICE 16: Access the definition of Environment in Linux.
DESCRIPCIÓN:

Linux provides ncal and cal utilities that can be used to display the online calendar of commands. Once you get used to them, you will notice that things are faster with these utilities compared to manually looking at the GUI calendars. Both utilities, when combined, offer a broad set of options through which the calendar can be displayed in almost any way.

STEP 1: Calendar.

There are two ways to visualize the calendar in columns and in rows with the orders:

```
cal
ncal
```

a)   Help.

```
cal  --help
ncal  --help
```

b)   Visualize the calendar of one year.

```
cal 2014
cal
August 2015
Su Mo Tu We Th Fr Sa
 1
 2  3  4  5  6  7  8
 9 10 11 12 13 14 15
16 17 18 19 20 21 22
23 24 25 26 27 28 29
30 31
```

c)   Display the calendar for a month.

```
cal mes  año
cal 6  2010
```

d)   Display by default.

```
ncal
Agosto 2015
lu      3 10 17 24 31
ma      4 11 18 25
mi      5 12 19 26
ju      6 13 20 27
vi      7 14 21 28
sá   1  8 15 22 29
do   2  9 16 23 30
```

e)   Calendar with horizontal display.

```
ncal 2014
```

f)   See the Easter lebration date.

```
ncal -e
```

5 abril 2015

```
ncal -e 2016
```

27 marzo 2016

g)   Display in sickle format. One month.

```
ncal 6 2013
```

h)   Various examples:

```
cal -m1
cal -m1 1968
cal -3 -m1
ncal -w
ncal -M
ncal -p
```

cal		
Syntax:	**cal [opcional] [-hjy] [[month] year]**	
	cal [opcional] [-hj] [-m month] [year]	
	ncal [opcional] [-bhJjpwySM] [-s country_code] [[month] year]	
	ncal [opcional] [-bhJeoSM] [year] Opcional : [-NC3] [-A months] [-B months]	
-1	Shows a single month as output.	
-3	Shows the previous/current/next month as output.	
-s	Show Sunday as the first day of the week.	
-m	Show Monday as the first day of the week.	
-j	Shows Julian dates (ordered days, numbered from January 1).	
-y	Shows a calendar for the current year	
ncal		
-h	Disable today's highlight	
-J	Show Julian Calendar, if combined with the option-e, Easter presentation date according to the Julian calendar.	
-e fecha	See the date of celebration of the Passover (for Western churches).	
-j	Display Julian days (day one-based, numbered January 1).	
-m mes	Shows the specified month. If the month is specified as a decimal number, it can be followed by the letter 'f' or 'p' to indicate the following or the month of that number, respectively.	
-o	Show date of the Orthodox Easter (Greek and Russian Orthodox Churches	
-p	Print the country codes and the commutation days of Julian the Gregorian calendar, since it is assumed by NCAL. The country code that is determined by the local environment is marked with an asterisk.	
-s cod_pais	Assume Julian's change to the Gregorian calendar, on the date associated with the country_code. If it is not specified, NCAL tries to guess the date of change of local environment or falls back to September 2, 1752 This was when Britain and its colonies changed to the Gregorian calend	
-w	Displays the week number below each column of the week.	
-Y	Show a calendar for the specified year.	

calendar [-ab] [-A num] [-B num] [-l num] [-w num][-f calendarfile] [-t [[[cc]yy][mm]]dd]	
-A	Number of future days.
-B	Number of previous days
-f	Calendar file.
-t	Value year 69 and 99
-w	Number of lines to be displayed (default 2).
-l	Visualize the lines of a number of future days.
-a	Process the calendar and send it to the mail of all users. Use in superuser mode.
-b	Force a calendar date in KOI8 mode.

STEP 2: Extract information about events that occurred on a specific date (Efemerides).

```
calendar
```

a)   Help.

```
calendar --help
```

b)   By default, the current date is assumed.

```
calendar
ago 17  Mae West born, 1892
ago 17  First public bath opened in N.Y., 1891
ago 17  Anniversary of the Death of General San Martin in Argentina
ago 17  Independence Day in Indonesia
ago 17  Odin's Ordeal
ago 17  Paso a la inmortalidad de José Francisco de San Martín, 1850
ago 17  Olivier Houchard <cognet@FreeBSD.org> born in Nancy, France, 1980
ago 17  Aujourd'hui, c'est la St(e) Hyacinthe.
ago 17  Jácint
ago 18  Meriwether Lewis born, 1774
ago 18  Anti-Cigarette League of America formed
ago 18  N'oubliez pas les Hélène !
ago 18  Temps trop beau en août
Annonce hiver en courroux.
ago 18  Ilona
```

ago 18 National Science Day (วันวิทยาศาสตร์แห่งชาติ) in Thailand

c) Efemerides of a specific date.

    calendar -t   2009-05-02

STEP 3: Date and time.

    date

a) Help.

    date --help
    man date

b) Change date.

    date   -s 2005-05-01
    date

c) Change the time.

    date -s 11:10
    date

d) Change the date and time.

    date -s "2012-05-28 10:41"
    date

e) Use a file to change the time.

    cat > fecha
    2012-05-28 11:30 ^D

f) Read the information from a file -f.

    date -f fecha

date	
Syntax:	**date [opciones][+format][date]**
-a	Slowly adjust the time in sss.fff seconds (fff represents fractions of a second). This adjustment can be positive or negative. Only the system administrator or superuser can set the time.
-sdate-string	Sets the date and time to the value specified in the datestring. The datestr may contain the names of the months, time zone, "am", "pm", etc.
-u	Displays (or sets) the date in Greenwich Mean Time (GMT-universal time).
Format:	
%a	Day of the week abbreviated (Tue).
%A	Day of the whole week (Tuesday).
%b	Name of the abbreviated month (Jan).
%B	Name of the full month (January).
%c	Time format and date specific to the country
%D	Date in %m /%d /% and format.
%j	Day of the Julian year (001-366).
%n	Insert a new line
%p	Chain to indicate a.m. or p.m.
%T	Time in%H:%M:%S format
%t	Tab space
%V	Number of the week in the year (01-52); start of the week on Monday.

STEP 4: View the time a user takes in a connection.

It shows the time, operating time, number of connected users and the average load.

    uptime

a) Help.

    uptime --help

b) Display by default.

    Uptime

STEP 5: Show the HW equipment clock, and its configuration.

    hwclock

a) Help.

    hwclock --help

b) Shows the Hardware clock or BIOS clock.

    hwclock --show
    hwclock -r

c) Asignar al reloj de Hardware la hora del sistema operativo.

    hwclock -systohc
    hwclock -w

d) Asignar al reloj del sistema la hora del reloj de la BIOS.

    hwclock --hctosys
    hwclock -s

e) Debugging.

    hwclock --hctosys --debug
    hwclock --systohc --debug

hwclock [funtion] [option]	
FUNCTION	**DESCRIPTION**
-r --show	Read the Hardware Clock and show the time in standard output.
--set	Put the Hardware Clock at the time given by the --date option
-s --hctosys	Set the System Time from the Hardware Clock.
-w --systohc	Sets the Hardware Clock to the current system time.
--adjust	Add or subtract time from the Hardware Clock to take into account the systematic deviation from the last time the clock was set or adjusted
--getepoch	Displays in the standard output the time value of the Core Hardware Clock.
--setepoch	Sets the time value of the Core Hardware Clock to the value specified by the --epoch option
--date=nuevafecha	Specify the time to put the Hardware Clock. hwclock --set --date = 9/22/96 16:45:05
--epoch=año	Specify the year that is the beginning of the Hardware Clock era. hwclock --setepoch --epoch = 1952
OPTION	**DESCRIPTION**
-u --utc	Indicates that the Hardware Clock is maintained in Coordinated Universal Time (UTC).

PASO 6: Ejecuta una orden cada cierto tiempo.

    watch

a) Help.

    watch --help

b) Execute an order every x seconds (default every 2 seconds).

    watch -n 5 hwclock -r

STEP 7: Clear the screen.

    clear

a) Help.

    clear --help
        man clear

The environment variables that it uses, UNIX, Linux, are all defined with uppercase names. (In the order in lowercase).
- Are defined, assignments in script language Shell (script), of directives or functions.
- The directives defined in the environment can be invoked or called at any time.

STEP 8: Environment or environment variables.

    set
    env

a) Help.

    set --help --> error: implies visualizing the help line

    man   set
b)   Display the environment variables.
        set
        set > salida
        less salida
c)   Variables only of user environment.
c.1.) Help.
        env --help
        help   env
c.2.) Variables only of user environment..
        env
        env > salida2
d)   Define an environment variable, the name of the variable is set equal to a value (variable = Value).
        HOLA=buenos días
d.1) Visualize the content of the concrete variable.
        set
d.2) Visualize the variables that begin with H.
        set H
d.3) Visualize the content of the concrete variable.
        echo $HOLA
        **
e)   Delete an environment variable.
        unset HOLA
      Variables always define parameters or routes. If they define routes, the routes are separated by (:).

VARIABLE	DESCRIPTION
PATH	Default route of alternative search.
HOME	User's work directory, (home)
PWD	Active route.
USER	Username.
SHELL	Shell of the interpreter of orders, carapace.
PS1	Parameters of the root prompt
PS2	Parameters of the users prompt.

UID: user identification, the user is identified by a number.
EUID: special user identification. Reference to rights.
Real rights. -> chmod, umask
Inherited rights.
Rights (special, implicit) Cash umask

        set
        env
        echo $PS2 $PS1

VARIABLE	DESCRIPCIÓN
MAIL	Route the file or mail files. Directory of the mail.
LOGNAME	Name of the connection or connection user.
HOSTNAME	Name of the team.
HISTFILE	Route and storage file of the historical, the set of orders written in a session or sessions.
HISTFILESIZE	Maximum number of open files of historical.
HISTSIZE	The maximum size per file or set of orders or entries that can be stored in the historical file.
BASH	/bin/bash the path and name of the command interpreter.
PPID	Identification of the parent process.

STEP 9: Acronyms or ALIAS.

   The alias command allows you to create a shortcut to a command. As the name indicates, you can set the name of the alias / shortcut for commands / routes that are too long to remember them.
        alias
a)   Help.
        alias --help
        man alias
b)   Visualize by default the aliases established in the system.
        Alias

alias	
Syntax:	**alias [options] [NameAlias [=String]]**
-a	Remove all alias definitions from the current Shell runtime environment.
-p	Show the alias list of the name name alias = value in the standard output.

c)   Define an alias The command is enclosed in single quotes (key? / ').
        alias   nombre= 'valor'
        alias   fichero='ls -la'
        alias   fich='ls –la |less'
        alias   busca='fichero|find   -name salida'
d)   Perform search execution and add a value to search and admit it with a replaceable parameter.
        alias buscar='find / -name '
        alias ll='ls -l'
        alias la='ls -A'
        alias l='ls -CF'

e)   Execute aliases predefined by the system.

      ll

      la

      l

    Other aliases:

      fich

      busca

      fichero

STEP 10: Delete acronyms or ALIAS.

Remove each name from the list of defined aliases.

      unalias

a)   Help.

      unalias --help

      man   unalias

b)   Delete by default

      unalias fichero

      unalias busca

Aliases prevail during the time that the session is active.

To use always you have to define the aliases in the personal file of the boot sequence.

> **Where do we put this?**
>
> Well, if we want it to be only temporary, we simply write it in the console and it will last until we close it.
>
> Now, if we want it permanently, we put it in the file ~/.bashrc which is in our /home, and if it is not, then we create it (always with the point in front). When we have added the alias line in this file, we simply put in console and execute:
>
>   **# . .bashrc**

> unalias
>
> **Syntax: unalias [-a] name [name ...]**
>
> | | |
> |---|---|
> | **-a** | Specify the name of the alias you want to delete |

WORK UNIT IV: General operations on Linux operating systems.

PRACTICE 17: Starting and stopping Linux.

PRACTICE 18: Boot levels, runlevel in Linux.

PRACTICE 19: Configure the network in Linux.

PRACTICE 20: Add applications or repositories in Linux.

PRACTICE 21: Configure the basic data of a UBUNTU server.

PRACTICE 22: Device information in Linux.

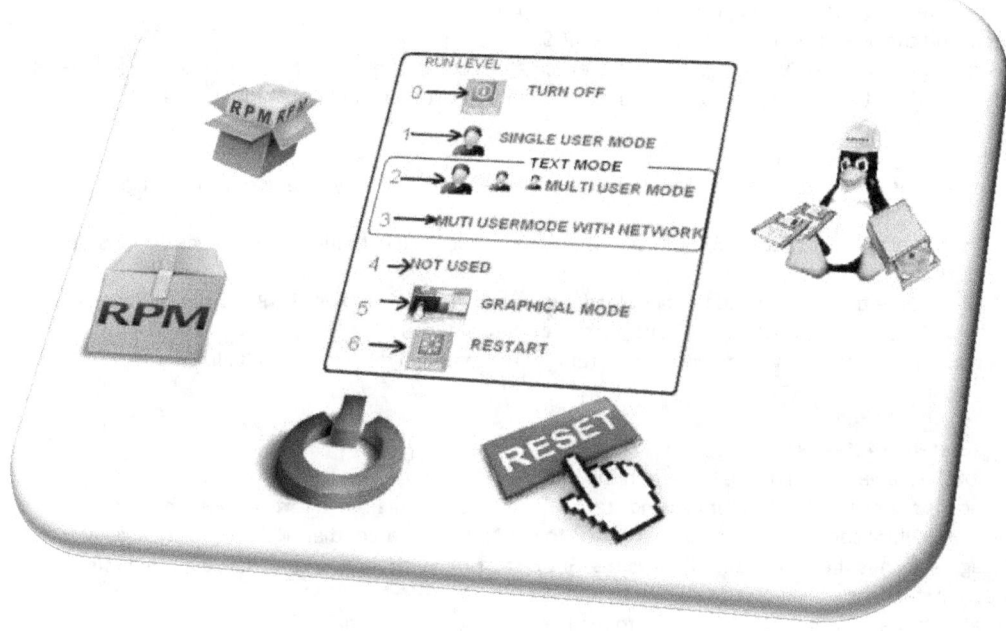

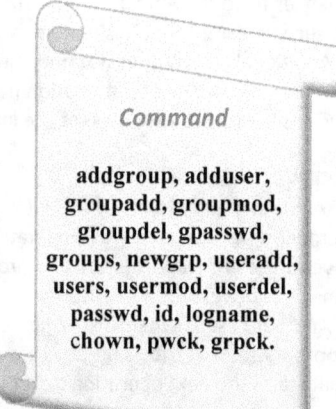

Command

addgroup, adduser, groupadd, groupmod, groupdel, gpasswd, groups, newgrp, useradd, users, usermod, userdel, passwd, id, logname, chown, pwck, grpck.

Content
- System administration.
- Groups in Linux.
- Users in Linux.

PRACTICE 17: Starting and stopping Linux.
DESCRIPTION:

Obtain information from the Linux machine and system, at the level of the boot sequence.

The boot process in GNU / Linux is the way in which operating systems based on the Linux kernel are initialized. It is similar to the way BSD and other Unix systems boot.

The whole boot process is carried out in 4 stages recognized by the code that at that moment has control over the CPU; at the beginning only the BIOS has control, then the boot loader will be in control, later the control goes to the Linux kernel itself, and in the last stage it will be when we have in memory the user programs coexisting together with the operating system itself and it will be they who have the control of the CPU.

The boot loader stage is not absolutely necessary, certain BIOS can load and pass control to GNU / Linux without using the boot loader, using a boot loader provides the user with the way the kernel will be loaded.

BIOS	Basic Input/Output System that identifies and executes MBR
M B R	Run from the BIOS (In the boot sequence) The Master Boot Record runs the GRUB
GRUB	Grand Unified Linux Bootloader executes the call to the Kernel
Kernel	The Kernel is an .img file in the / boot and the executables are / sbin / init
init	init is executed at the level of the runlevel, and access to them leaves
Runlevel	Executable runlevel programs are located in /etcrc.d/rc*.d

1. BIOS.

When the computer is turned on, the first operations are performed by the BIOS. In this stage, basic hardware operations are performed. The boot process will be different depending on the processor architecture and the BIOS.

Once the hardware is recognized and ready, the BIOS loads the executable code of the boot loader into memory and passes the control to it. There are a variety of BIOS that allow the user to define in which device / partition that boot loader is located.

2. Starter charger.

A boot loader is a program designed exclusively to load an operating system into memory. The boot loader stage is different from one platform to another.

As in most architectures, this program is in the MBR, which is 512 bytes, is not enough to fully load an operating system. Therefore, the boot loader consists of several stages.

For x86 platforms, the BIOS loads the first stage of the boot loader (typically a part of LILO or GRUB). The code of this first stage is in the boot sector (or MBR). The first stage of the boot loader loads the rest of the boot loader.

Modern boot loaders typically ask the user which operating system (or session type) they want to initialize.

2.1. GRUB

GRUB is loaded and executed in 4 stages:
1. The first stage of the loader is read by the BIOS from the MBR.
2. The first stage loads the rest of the charger (second stage). If the second stage is in a large device, an intermediate stage (called stage 1.5) is loaded, which contains extra code that allows to read cylinders larger than 1024 or LBA-type devices.
3. The second stage executes the loader and displays the GRUB start menu. Here you can choose an operating system together with system parameters.
4. When an operating system is chosen, it is loaded into memory and the control is passed.

GRUB supports direct boot methods, chain-loading boot, LBA, ext2 and even "a fully command-based pre-operating system". It has three interfaces: a selection menu, a configuration editor and a command line console.

Since GRUB understands the ext2 and ext3 file systems and also provides a command line interface, it is easier to rectify or modify when it is misconfigured or corrupted. The new version 2 of GRUB, supports ext4 file system.

2.2. LILO

LILO is older, it is almost identical to GRUB in its process, except that it does not contain a command line interface. Therefore, all changes to your configuration must be written to the MBR, and restart the system. An error in the configuration can ruin the boot process to such an extent that it is necessary to use another device that contains a program that is able to fix that defect.

Additionally, LILO does not understand the file system, therefore there are no files and everything is stored in the MBR directly.

When the user selects an option from the load menu of LILO, depending on the response, it loads the 512 bytes of the MBR for systems such as Microsoft Windows, or the image of the Linux kernel.

Loadlin

Another way to load GNU / Linux is from DOS or Windows 9x, since both systems allow to be replaced, it can be replaced by the Linux kernel on the already loaded operating system. This can be useful in case the hardware is only available for DOS and not for GNU / Linux, given issues of trade secrets and proprietary code. However, this tedious way of starting is no longer necessary at present since GNU / Linux has drivers for a multitude of hardware devices, even so, this was very useful in the past.

Another case is when GNU / Linux is on a device that the BIOS does not have available for boot. Then, DOS or Windows can load the appropriate driver for said device overcoming this limitation of the BIOS, and thereafter load the Linux kernel. If you have the .img file that starts the next operating system.

Kernel

The Linux kernel is responsible for the main operating system, such as memory management, task scheduler, inputs and outputs, interprocess communication, and other control systems.

The kernel process takes two stages; the loading stage and the execution stage.

The kernel is usually stored in a compressed file with zlib. This compressed file is loaded and decompressed in memory, also the necessary drivers are loaded by means of a RAM disk (initrd). The RAM disk is a temporary file system used in the execution phase of the kernel.

Once the kernel has been loaded into memory and is ready, execution is carried out. This is done by calling the kernel startup function (in the x86 processors, it is found in the startup_32 () function of the file /arch /i386/boot/head), this function establishes the memory management (paging tables and memory paging)), detects the CPU type and additional functionality as floating point capabilities. Then it changes to functionalities that do not depend on the hardware by means of the call to the function start_kernel ().

The boot process in GNU / Linux mounts the RAM disk that was previously loaded as a temporary file system. This allows the modules containing drivers to be loaded without relying on other physical device drivers, and also keeps the kernel smaller.

Virtual devices are initialized with the intention of being used to create file systems, such as LVM or software RAID before disassembling the initrd image. The file system is changed by means of the pivot_root () function, which unmounts the temporary file system and replaces it with the real one, which will later be fully available, freeing the memory occupied by the storm.

Once the exception handler, the task scheduler and so on are ready, finally the system is considered fully operational at the process level, therefore the init process is executed (the first process in user space), and then initiates a inactivity task by means of cpu_idle ().

Process init

The init process establishes the user environment. Verify and mount file systems, start required user services, and switch to a user-based environment when the startup process ends.

It is similar to the init processes of Unix and BSD from which it derives, but in some cases it has differences and customizations. In a standard GNU / Linux system, init is executed with a parameter, known as runlevel, which takes a value from 0 to 6, and which determines which subsystems will be operational.

Each runlevel has its own scripts which involve a set of programs. These scripts are stored in directories with names like "/etc/rc ...". The init configuration file is /etc/inittab.

When the system is started, it is verified if there is a default runlevel in the file /etc/inittab, if not, it must be entered through the system console. Then we proceed to execute all the scripts related to the specified runlevel.

STEP 1: Information of the S.O.

    uname

a)   Help.

    uname --help

b)   Default.

    uname

    What do you drive?

c)   Display all the information of the S.O.

    uname -a

d)   Version of the operating system and kernel system.

    uname -s

e)   Last revision of the version.

    uname -r

f)   Kernel and the revision.

    uname -r -s

g)   Operating system version.

    uname -v

h)   The node within a network or else the name of the machine.

    uname -n

i)   Type of microprocessor is in this machine.

    uname -m

j)   Type of processor within the family.

    uname -p

uname	
Syntax:	**uname [option] ...**
OPTION	**DESCRIPTION**
-a	Visualize all the information, in order:
-s	Visualize the name of the kernel, just as without the default option.
-n	Display the host name of the network node.
r	Shows the kernel version
v	Visualize the kernel version
-m	Display the hardware name of the machine.
p	Visualize the type of processor.
i	Visualize the hardware platform.
-o	Visualize the operating system.

STEP 2: Establish the connection of a user.

The connection can be made:

    login

    su

a)   In a text console.

    login

a.1) Help.

    login --help

a.2) Default connection.

    login

a.3) Connection conserving the environment.

    login -p

a.4) Connection with the name of a remote computer (computer name: position-01).

    login -h puesto-01 -f alumno

a.5) Connection with the name of a remote computer (computer name: position-01).

    login -p -f alumno

login	
syntax:	**login [-p] [user]**
	login [-p] [-h host] [-f nombre]
	login [-p] –r host
OPTION	**DESCRIPTION**
-p	Conserve the environment
-h host	Login remote computer connection.
-r host	Remote computer connection rlogin
-f user	The user is preautoconfigured.

b)   Connect with an application eg Putty, using the SSH service.

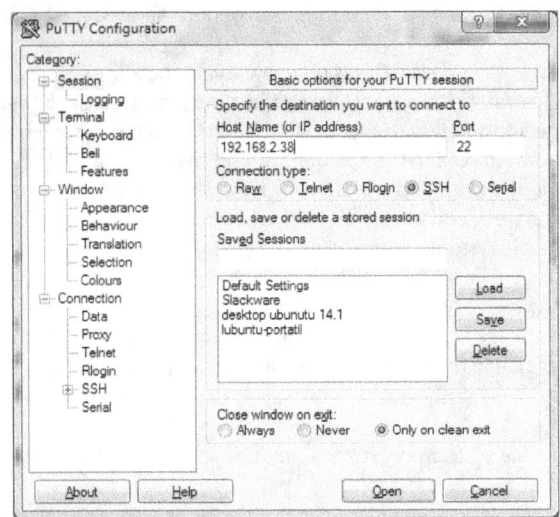

To be able to manage an application, which uses the SSH protocol, you must first configure the SSH service, its configuration is in practice.

Download address of **PUTTY**
http://www.putty.org/

c) Connect with the application.

Initially it is at the level of a user account eg: student, it should not be at the root level. This is well defined Ubuntu, does not allow the connection at the root level, but user, once the connection you can change user to root or other.

STEP 3: Close a connection.
        exit
a)   Help.
        exit --help
      Going out, has left without showing the help, ...
b)   Help with man.
        man exit

STEP 4: Exit with open logout.
        logout
a)   It belongs to layer 6: SESSION.
        logout   Close the connection
        login   establish a new connection
b)   Close connection.
        logout
c)   Open a new connection.
        login   --> user y password
d)   Request only the passwd of a connection user.
        login   alumno3
e)   A new session opens.
        exit
        login alumno3
        logname
        logout -->   does not allow in this case to close the connection
        exit
        logout -->   last connection which closes.
f)   Can you consider that your is opening a new connection?
        su   alumno10

The first user that makes the connection to a Linux by ssh, starting from kernel 2.6 can not be the root. The initial connection is with normal user and once inside you can change the connection user, to root.

last	
Syntax:	**last [options]**
OPTION	**DESCRIPTION**
-n	Specify how many lines to show.
-R	It does not show the hostname field.
-x	It shows the shutdown entries of the system and the changes in the execution levels.
-a	Shows the hostname in the last column. Useful in combination with the following flag.

STEP 5: Show the last users connected to the system.
        last
a)   Help.
        last --help
b)   Default.
        last
        last > usuarios5
c)   Displays the number of lines to display 5.
        last -n
d)   Do not display the hostname field.
        last -R
e)   It shows the entries made during system shutdown and level execution changes.
        last -x
f)   Display the hostname in the last column.
        last -a

STEP 6: Show the last users who have tried to connect.

      Lastb

a)   Help.

      lastb --help

b)   Default.

      lastb

      lastb -d

      lastb -f

      lastb -oar

STEP 7: Date and time of the last login made by the active user.

The lastlog command is used to show the last connection time of the system accounts. The access information is read from the /var/log/lastlog file.

      lastlog

a)   Help.

      lastlog --help

b)   Default.

      lastlog

c)   Display the lastlog record with the values of the last 5 days.

      lastlog -b 5

d)   Latest information.

      lastlog -t 10

e)   Login of a specific user.

      lastlog -u root

      lastlog -u alumno10

      lastlog -u alumno3

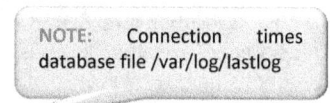

NOTE: Connection   times database file /var/log/lastlog

lastlog	
Syntax:	**lastlog [options]**
OPTION	DESCRIPTION
-t n	Show only accesses for less than "n" days ...
-u name_user	Show only the login information for the username.

PRACTICE 18: Boot levels, runlevel in Linux.
DESCRIPTION:

Runlevel

The runlevel (of English, level of execution) is each one of the states of execution in which the Linux system can be found. There are 7 levels of execution in total: (Maximum 9, of which 7 are defined).

- Execution level 0: Off.
- Execution level 1: Single user (only root user, password is not necessary). It is usually used to analyze and repair problems. (boot: init 1)
- Execution level 2: Multiuser without network support.
- Execution level 3: Multi-user with network support.
- Execution level 4: As the runlevel 3, but it is not usually used.
- Execution level 5: Multi-user in graphic mode (X Windows).
- Execution level 6: Restart.

> Ubuntu is different. The /etc/inittab file was replaced as of version 6.10 by /etc/upstart
> Which now has also changed. And the levels of execution are the following:
> **0 - shutdown**
> **1 - single-user mode**
> **2 - single-user graphic mode**
> **6 - reboot**
> In /etc/init/rc.conf (see Ubuntu 16.04)

This system of execution levels is provided by the default boot system of most GNU / Linux distributions (init). However, Canonical has been developing a new startup system called upstart to replace init, since init is not adapted to current needs.

Modify the runlevel by default

By default, the system usually starts at execution level 5 (graphic mode). If you want to modify this behavior, you should edit the file /etc/inittab.

id: execution_layers: action: process

More specifically, the line should be modified in the /etc/inittab file.

id:5:initdefault:

Where the number 5 indicates that the default execution level is 5. This number is the one that must be modified to change the execution level at which the system starts by default.

        cat   /etc/inittab
Check the current execution level?

        who -r
        'run-level' 2 2014-09-08 12:12
        runlevel
        N 2

STEP 1: Turn off the equipment.

There are different ways to shut down the Linux system, but all refer to init 0.

        init 0
        halt
        poweroff
        shutdown

a)   Help.

        halt --help

b)   Advisable.

        halt -p

c)   Timer the shutdown and send messages to users connected shutdown.

c.1.) Help.

        shutdown --help

c.2.) Turn off with the previous sending of a message, notification "The equipment will be shut down shortly".

        shutdown -k   The equipment will be shut down shortly -c 30 –h

STEP 2: Restart the system.

There are different ways to restart the Linux system, but all refer to init 6.

        init 6
        reset
        halt -w
        shutdown -r
        reboot

STEP 3: Unit synchronization storage units and buffer.

Synchronization writes the data temporarily stored in the memory to the disk. This may include (but is not limited to) modified superblocks, modified inodes, and delays in readings and writes.

        sync

a)   Help.

        sync --help

> Required to use it in disk raid, LVM partitions.

b)   Default.

        sync

STEP 4: Start in text mode and move to the graphical environment.

If the system was started in runlevel 3, and we want to go to runlevel mode 5.

startx

To start or lift the graphic mode (xinit).

ORDER	DESCRIPTION
startx -- :DISPLAY	Start the graphical server indicating the DISPLAY, by default the first DISPLAY is 0 (which we access with Cntrl + Alt + F7).
startx -- :1	We start the graphic server in DISPLAY 1, that is to say in the console that is accessed by Control + ALT + F2, in which after pressing that combination of keys (Control + ALT + F2) and start transfer we can execute graphic applications.
X :DISPLAY	If we do not want to start any desktop and we want to use the X (run graphic applications without lifting GNOME).
X :3	Start the fourth graphic server. Most of the graphic programs in GNU / Linux support the -display or-display option with which the graphical server is indicated where we want it to run.

Having started the X in a display: 3 it can be useful to take out a console in order to execute things like: gnome-terminal-display: 3.

Go to DISPLAY: 3 through the combination of keys Cntrl + Alt + F10 we will see an Xterm where you can execute commands, either to start a desktop or another application such as a game.

If we want to start the X together with a console we can use the xinit command, which by default starts an Xterm.

STEP 5: Display or change runlevel information.

It is used to change, update and query runlevel information for system services. The chkconfig command is administrator.

chkconfig

a) List the execution levels and the service status.

chkconfig --list

b) The previous configuration command lists the execution levels and the status of the service (whether it is active or not).

chkconfig tomcat5 off

c) The above command is used to set the status for the tomcat5 service. Now the status of the tomcat5 service is inactive.

chkconfig --list tomcat5

PRACTICE 19: Configure the network in Linux.

DESCRIPTION:

Configure the network card. Can be done:

- d) In graphic environment.
- e) From the text console.
  - Configuration files.
  - Order line.

The configuration is explained from the console, in the configuration files, you can change from one version to another.

To the configuration file of the network card.

/etc/init/networking.conf	--> start the boot parameters.
/etc/resolv.conf	--> Resolution of the DNS (PRIMARY, SECONDARY)
/etc/network/interfaces	--> Configuration of network cards.

auto lo
iface lo inet loopback

auto eth0
if ace eth0 inet static

address	192.168.0.170
netmask	255.255.255.0
network	192.168.0.0
broadcast	192.168.0.255
gateway	192.168.0.100f

> The configuration of the network card is mandatory if we want to be able to download, the applications and update the repositories, as well as the configuration of the SSH service, to be able to use remote administration.

> Ubuntu change the names of network devices, from v. 16.04.
> The name of my network interfaces in the following way:
> - enp3s0 previously: eth0, eth1, …
> - wlp1s0 previously wlan0, wlan1, …

cat /etc/resolv.conf
ifconfig
cat /etc/network/intefaces

STEP 1: Modify the file / etc / network / interfaces.

nano /etc/network/interfaces
ctrl+x --> save
¿ yes ?   Y [enter]

STEP 2: Name of the team.

/etc/hostname
nano /etc/hostname
ctrl+x
record **Yes**.
with the name hostname [**ENTER**].

STEP 3: Modify the DNS resolution file.

/etc/resolv.conf
nano /etc/resolv.conf

STEP 4: Configure on the order line or PROMPT.

The ifconfig command is used, an address and a mask are assigned to the network device.
ifconfig eth0   192.168.2.197   netmask 255.255.255.0
Assign an alias to a network card.
ifconfig eth0:1 192.168.2.198   netmask 255.255.255.0

STEP 5: Restart the changes /etc/init.d/

File to run networking.
cd /etc/init.d
ls -l networking
Execute the file.
./networking   restart
. networking restart
If I am in root mode, it expels me to user mode.

> Know and change the MAC addresses of a network card PERMANENT. We edit the following file.
>   nano /etc/udev/rules.d/10-network.rules
> The following values are modified:
> SUBSYTEM=="net", ACTION=="add", ATTR(address)=="00:11:xx:xx:xx:xx", NAME=="eth0"

STEP 6: Change our network interface temporarily.

We turn off our network interface by executing the following command in the terminal.
sudo ifconfig enp3s0 down
Then we change the name of the interface from enp3s0 to eth0 by executing the following command in the terminal.
sudo ip link set enp0s3 name eth0
Finally we raise the new network interface eth0 by executing the following command.
sudo ifconfig eth0 up

PRACTICE 20: Add applications or repositories in Linux.
DESCRIPTION:

What is a repository of Linux Debian or Ubuntu applications?

A repository consists of at least one directory with some DEB packages in it, and two special files that are **Packages.gz** for binary packages and **Sources.gz** for source packages.

Once your repository is listed correctly in **sources.list,** if the binary packages are listed with the keyword deb at first, apt **Packages.gz** search the index file, and if the sources are listed with keywords Deb src at the beginning, it will search the **Sources.gz** index file.

This is because in the **Packages.gz** file all information of all packages, such as name, version, size, short description and long, dependencies and any additional information that is not of our interest is. All information is listed and used by the System Packages Administrators such as dselect or aptitude.

However, in the **Sources.gz** file you are listed all the names, versions and development units (packages are needed to compile) of all packages whose information is used by **apt-get source** or similar tools.

Once you have established your repositories, you will be able to list and install all their packages next to those that come in the Debian installation disks; Once you have added the repository you must run in the console:

> # *aptitude update*

This is in order to update the database of our APT and so he can "tell us" which packages we have with our new repository. The packages will be updated when we run in console.

> # **aptitude upgrade**

We will use apt-get to add packages.

There is a repository found. /etc/apt

> ls -l /etc/apt

sources.list

It contains the references of the repository servers, it is a text file.

Notification in the connection or access to the console terminal.

```
Welcome to Ubuntu 15.04 (GNU/Linux 3.19.0-15-generic x86_64)
Documentation:  https://help.ubuntu.com/

7 packages can be updated.
7 updates are security updates.
```

STEP 1: Edit the file /etc/apt/sources.list

a)   Access the directory

> cd   etc/apt
> ls -l sources.list

b)   Edit the sources.list file, to add the update of the repositories:
- Special files: deb
- The binary packages: deb-src

> # nano sources.list

STEP 2: Add certain orders, applications ...

a)   Add the order tree.

> apt-get   install tree

If it does not work, you need to review the following steps:

a.1) Previous step to install, you have to resolve DNS.

> nano /etc/resolv.conf
> search 192.168.0.100   80.58.61.250   80.58.61.254   8.8.8.8
> nameserver 80.58.61.250
> nameserver 80.58.61.254
> nameserver 8.8.8.8
> cd /etc/init.d

> **NOTE:** Help man will appear by default in English

a.2) Restart the network service, to load the new values.

> ./networking restart

a.3) Check the configuration and install.

> ifconfig
> apt-get   install tree

b)   Add a GNOME kernel.

> apt-get install xorg gnome-core

c)   Install language applications (Spanish):

> apt-get install language-pack-es
> apt-get install language-pack-es-base

d)   Install the language pack for the gnome.

> apt-get install language-pack-gnome-es
> apt-get install language-pack-gnome-es-base

e)   Install a selector.

> apt-get install language-selector
> apt-get install language-support-es

STEP 3: Update versions.
a) Make an update of the version of our system, to make sure that it is in the most recent, using the following order line:
    ```
    sudo do-release-upgrade
    do-release-upgrade
    ```
b) It is advisable or convenient to update to the latest version of the different packages that are installed:
    ```
    sudo apt-get update && sudo apt-get -y dist-upgrade
    apt-get update && sudo apt-get -y dist-upgrade
    ```

STEP 4: Install the graphical environment in UBUNTU SERVER.
```
sudo apt-get install Ubuntu-desktop
```
a) After the execution of this instruction we will have installed the Gnome graphical environment in its entirety, which includes a lot of desktop tools that are not normally needed on a server, such as Libre Office, and that also consume resources. To avoid this, it is possible to use a second alternative, which only installs a minimum desktop configuration:
    ```
    sudo apt-get install x-window-system-core gnome-core
    ```
b) After installation, to start the graphical environment, execute the following:
    ```
    startx
    ```
c) To configure the language in Spanish it will be necessary to install the following packages:
    ```
    sudo apt-get install language-pack-es
    sudo apt-get install language-pack-es-base
    sudo apt-get install language-pack-gnome-es
    sudo apt-get install language-pack-gnome-es-base
    sudo apt-get install language-selector-gnome
    ```

APLICATION	CONFIGURATION OF THE REPOSITORY
Medibuntu	deb http://packages.medibuntu.org/ intrepid free non-free Install the medibuntu-keyring and app-install-data-medibuntu packages.
Wine	deb http://wine.budgetdedicated.com/apt intrepid main the wine package
OpenOffice.org 3.0	deb http://ppa.launchpad.net/openoffice-pkgs/ubuntu intrepid main
Opera	deb http://deb.opera.com/opera/ stable non-free and install the opera package.
Banshee	deb http://ppa.launchpad.net/banshee-team/ubuntu intrepid main and we will install the banshee package
VideoLAN Client (VLC)	deb http://ppa.launchpad.net/c-korn/ubuntu intrepid main Once the source list is updated, we only need to install the vlc package
Boxee	deb http://apt.boxee.tv intrepid main The package to install, boxee
Elisa	deb http://ppa.launchpad.net/elisa-developers/ppa/ubuntu intrepid main and the package to install, as expected, elisa
Netbook Remix	deb http://ppa.launchpad.net/netbook-remix-team/ubuntu intrepid main To be able to enjoy UNR it is necessary to install the go-home-applet, human-netbook-theme, maximus, netbook-launcher and window-picker-applet packages and run the netbook-launcher and maximus at the start
Gnome Do	deb http://ppa.launchpad.net/do-core/ppa/ubuntu intrepid main
Deluge	deb http://ppa.launchpad.net/deluge-team/ubuntu intrepid main and the name of the package to install, deluge
Google Gadget	deb http://ppa.launchpad.net/googlegadgets/ppa/ubuntu hardy main The package we are interested in is google-gadgets
Mythbuntu	deb http://ppa.launchpad.net/mythbuntu/ubuntu hardy main
Compiz	deb http://ppa.launchpad.net/compiz/ubuntu intrepid main and update the system.
Miro	deb http://ftp.osuosl.org/pub/pculture.org/miro/linux/repositories/ubuntu intrepid/ and the package to install, I look
Mundo geek	deb http://ppa.launchpad.net/zootropo/ppa/ubuntu intrepid main

Firmas GPG
The GPG signatures of the repositories that require them, in case the package manager complains:

APLICATION	URL of GPG signatures for repositories
OpenOffice	http://keyserver.ubuntu.com:11371/pks/lookup?op=get&search=0x60D11217247D1CFF
Gnome-DO	http://keyserver.ubuntu.com:11371/pks/lookup?op=get&search=0x28A8205077558DD0
Deluge	http://keyserver.ubuntu.com:11371/pks/lookup?op=get&search=0xC5E6A5ED249AD24C
Google	https://dl-ssl.google.com/linux/linux_signing_key.pub
WineHQ	http://wine.budgetdedicated.com/apt/Scott%20Ritchie.gpg

PRACTICE 21: Configure the basic data of a UBUNTU server.

DESCRIPTION:

Configuration of the network card.

ifconfig -> Allows you to view configuration data and establish configuration data.

- eth0, enp3s0: Network card, the first of the network cards. Ethernet
- lo: localhost, 127.0.0.1

  ping   localhost

  ping   127.0.0.1

STEP 1: See the network configuration.

ifconfig

192.168.2.245

According to CIDR, the number of bits that make up the masks are indicated by ir / n_bits, the n_bits depends on the categories of the masks: a, b, c and others.

8 -> a

16-> b

24 ->c

a)   Help.

ifconfig --help

b)   Defect, shows the information of the network cards.

ifconfig

c)   Set an IP address and mask, tests or temporarily.

ifconfig eth0 192.168.0.150 netmask 255.255.255.0

ifconfig

ifconfig eth0:1 192.168.0.180 netmask 255.255.255.0

ifconfig

> NOTE: it is used with the route command, to perform routing tables..

STEP 2: Team name.

hostname

a)   Help.

hostname   --help

b)   Default.

hostname

c)   Display the domain to which it belongs.

hostname -d

d)   See all addresses IPs..

hostname -I

e)   View the assigned IP address.

hostname -i

f)   See the server assigned by default, boot

hostname -b   (boot)

hostname

Syntax: hostname [Options]

OPTION	DESCRIPTION
-a	Displays the host alias, if it exists.
-d	Show the DNS domain name
-f	It shows the fully qualified domain name.
-h	Show help messages
-i	Displays the host's IP address.

> NOTE: El resto de las opciones se utilizan con los servidores de dominio, especialmente con el protocolo LDAP--> Active Directory.

STEP 3: See the configuration of the Team at the version and core level of the S.O.

w

a)   Help.

man w

w   --help

b)   Default.

w

c)   Display version.

w -V

d)   Remove the title of the header.

w -h

e)   Ignore the identification of processes by your U

w -u

w

Syntax: w [-husfV] [user]

OPTION	DESCRIPCION
-h	He does not write the header.
-u	It does not take into account the username when the time of the current and CPU process is checked. To show this, make a "su" and make a "w" and a "w -u"
-s	Use the short form. It does not write the connection time, nor JCPU, nor PCPU.
f	Change the writing of the from field (name of the remote node). By default, the from field is not written, but the administrator of your system or the distribution supervisor may have compiled a version in which the from field is displayed by default.
-V	Shows information about the version.
usuario	Displays only information about the specified users.

STEP 4: View the connected terminals.

Displays serial devices, virtual terminals (accessible in order with the keys Alt-F1 to Alt-Fnn in the local console), "/dev/pts/0", which designates that the user is using the device "/pts/0", But it does not show any information about this device.

tty

stty

a)   Help.

tty --help

b)   Default.

tty

c)   Help.

stty   --help   // Terminals series

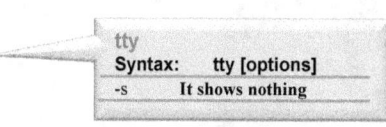

tty

Syntax: tty [options]

-s	It shows nothing

d) Default.

    stty

/etc/utmp information about who is connected at the time.
/proc information about processes.

STEP 5: Who I am.

    who a im
    whoami

a) Help.

    whoami --help

b) User name.

    whoami

STEP 6: Display information of the users who are currently connected.

The who command can list the names of the currently connected users, their terminal, the time they have been connected, and the name of the host from which they have connected.

    who

a) Help.

    who   --help
    man who

b) Default.

    who

c) System start date and time.

    who -b

d) Execution level.

    runlevel 0-6   --->   init <runlevel>
    who -r

e) Visualize dead processes

    who -d

f) Visualize the connection initiation processes.

    who -l

g) Display all the information.

    who -a

who	
Syntax:	**who [options] [archive]**
OPTION	DESCRIPTION
am i	Displays the user name of the person who invokes it. The "am" and the "i" must be separated.
-b	Shows the time of the last system boot.
-d	Show the dead processes.
-H	Show the column headings above the output.
-i	Include time stopped as HOURS: MINUTES. A time stopped and indicates activity at the last minute
-m	Same as who am i.
-q	Show only the user names and the active user account.
-T,-w	Include the user's status message in the output.

STEP 7: Displays the numeric identifier of the current host in hexadecimal.

    hostid
    root@192:~# hostid
    0000c000

WORK UNIT V: System Administration I. Network configuration. Administration of users and groups

PRACTICE 22: Manage groups in Linux.

PRACTICE 23: Manage users in Linux with orders.

PRACTICE 24: Manage users and groups in Linux with Perl scripts.

Content:
- Configuration start and end of session.
- Disk management in Linux.
- Update of the Operating System.
- Manage operating system hardware.
- Monitoring and system performance.
- Add/Delete/Update software in the operating system.
- Task programming in Linux.

PRACTICE 22: Manage groups in Linux.
DESCRIPTION:

The administration of a Linux operating system, requires a correct installation and configuration, then the system must start without errors.

From this point begins the tasks of planning the start and stop of the system, system monitoring, safety copies, options on groups and users: create, modify and delete, etc ...

STEP 1: Create new group accounts.
        groupadd
a) Help.
        groupadd --help
b) Create a group without specifications.
        groupadd   toreros
c) Create a group by assigning an identification.
        groupadd -g 1300 cuadrilla
    The following line is found in the /etc/group file.
        root:x:0:
        root -> group name
        x -> the key is in the file / etc / gshadow
        0 -> GID or GUID
        Blanco -> users who are part of the group.
        groupadd   prensa
d) Create a group of devices (-r).
        groupadd -r -g 197 usbdata
e) Create a group with the same id of another group (-o).
        groupadd -o -g 1300   ciclos

groupadd

Syntax: groupadd [options]   group_name

OPTION	DESCRIPTION
-f	Terminates if the group already exists, and cancels -g if the GID is already in use.
-g	Use GID for the new group.
-h	Show this help message and it ends.
-K Key=VALUE	Overwrites the default values of /etc/login.def
-o	It allows creating groups with duplicate GIDs (not unique)
-p PASSWORD	Use this encrypted password for the new group.
-r	Create a system account.
-R CHROOT_DIR	Directory in which to do chroot.

> The next GUID is from the last number assigned, previously, initially it is 1000.

STEP 2: Modify the parameters of a group.
        groupmod
a) Help.
        groupmod --help
b) Modify the identification of a group.
        groupmod -g 1350 toreros
c) Modify or change the identification to coincide with another group.
        groupmod -o -g 1350 ciclos
d) Modify the gid of a device.
        groupmod -g 297 ciclos
e) Change the name to a group.
        groupmod -n ies ciclos

groupmod

Syntax: groupadd [options] group_name

OPTION	DESCRIPTION
-g GID	Change the group ID to GID.
-n GROUP_NEW	Change the name to GROUP_NEW.
-o	Allows you to use a duplicate (not unique) GID
-p PASSWORD	Change the password to PASSWORD (encrypted)

STEP 3: Delete groups.
        groupdel
a) Help.
        Groupdel --help
b) Delete by default.
        groupadd nuevo
        cat /etc/group
        groupdel nuevo
        cat /etc/group
c) Delete the directory.
        groupdel -R

groupdel

Syntax: groupdel [options] group_name

OPTION	DESCRIPTION
-R ROOT_DIR	Directory in which to do chroot

gpasswd

Syntax: groupadd [options]   group_name

OPTION	DESCRIPTION
-a USUARIO	Add USER to the GROUP.
-d USUARIO	Remove USER from the GROUP.
-Q CHROOT_DIR	In the Chroot Directory.
-r	Remove the GROUP password.
-R	Restricts GRUPO access to its members.
-M USUARIO,...	Sets the list of GROUP members.
-A ADMIN,...	Sets the list of GROUP administrators.
Excepto las opciones -A y -M, las opciones no se pueden	

STEP 4: Set the password to a group.
        gpasswd
a) Help.
        gpasswd --help
b) Set password to a group by default.
        groupadd -g 1350 torero
        groupadd -g 1400 informatica
        groupadd -g 1500 bachillerato
        groupadd -g 1600 eso

        gpasswd torero
        passwd...:
        repetir....:
        useradd   -m -d /home/user1   user1

> The password of a group is used to set group changes to a user. Ask for the password.

```
        useradd   -m -d /home/user2   user2
        useradd   -m -d /home/user3   user3
```
c)   Add users to a group.
```
        gpasswd  -a user1  torero
        gpasswd  -a user2  torero
        gpasswd  -a user3  torero
        cat  /etc/group
```
d)   Delete users from a group.
```
        gpasswd  -d user3  torero
        cat  /etc/group
        gpasswd  -d user2  torero
        gpasswd  -d user1  torero
        cat  /etc/group
```
e)   Add a list of members to a group.
```
        gpasswd  -M  user1,user2,user3   torero
        cat /etc/passwd
```
f)   Assign a user as group administrator.
```
        gpasswd  -A  user1  torero
```
g)   Delete the key of a group.
```
        gpasswd  -r   torero
```

Groups predefined by the system

Group ID	GID
bin	1
sys	3
adm	4
tty	5
disk	6
lp	7
mem	8
kmem	9
wheel	10
mail	12
man	15
floppy	19
named	25
rpm	37
xfs	43
apache	48
ftp	50
lock	54
sshd	74
nobody	99
users	100

STEP 5: See the group to which a member belongs, user.

It shows the group membership of each USERNAME, or, if no USERNAME is specified, the current process (which may be different if the group database has changed).
```
        groups
```
a)   Help.
```
        groups  --help
```
b)   Create a default group, without options.
```
        groups  user1
```

STEP 6: Specify the default group of a user.

It specifies which is the default group of a user, the default group is used for example to specify the group of a new created file.
```
        newgrp
```
a)   Help.
```
        newgrp  --help
```
b)   Assign by default the current user to the group by default of users (by default it is users, or the name of the group that is named as the user, but it was indicated in its creation that it would be by default.
```
        newgrp
```

System groups

root	Owner of most system files.
daemon	Owner of the mail, printer and other system software and directories.
kmem	Manage direct access to kernel memory.
sys	Owner of system files, swap files, and memory images.
nobody	Software owner without special permits.
tty	Files of devices that control the terminals.
users	System users

PRACTICE 23: Manage users in Linux with orders.

DESCRIPTION:

What we can do with users:

- Register users.
- Modify users.
- Delete users.
- See the file of keys and restrictions.
- Set password.
- Display the identification of a user.
- Block user accounts.
- Unlock user accounts.

PERMISSION	DESCRIPTION
r	The permission to read a file.
	The permission to read a directory (also requires "x").
w	The permission to delete or modify a file.
	The permission to delete or modify files in a directory.
x	The permission to execute a file / script.
	The permission to read a directory (also requires "r")
s	User ID or group in the Set execution.
u	Permissions granted to the user who owns the file.
t	Set "sticky bit." Execute file / script as root user for the normal user.

Types of Users

- Also called supe user or administrator.
- Your UID (User ID) is 0 (zero).
- It is the only user account with privileges over the entire system.
- Full access to all files and directories regardless of owners and permissions.
- Controls the management of user accounts.
- Executes system maintenance tasks.
- You can stop the system.
- Install software in the system.
- Can modify or reconfigure the kernel, drivers, etc..

Also called superuser or administrator

- Your UID (User ID) is 0 (zero).
- It is the only user account with privileges over the entire system.
- Full access to all files and directories regardless of owners and permissions.
- Controls the management of user accounts.
- Executes system maintenance tasks.
- You can stop the system.
- Install software in the system.
- Can modify or reconfigure the kernel, drivers, etc.

Special users

- Examples: bin, daemon, adm, lp, sync, shutdown, mail, operator, squid, apache, etc.
- They are also called system accounts.
- It does not have all the privileges of the root user, but depending on the account they assume different root privileges.
- The above to protect the system from possible ways of violating security.
- They do not have pa sswords because they are accounts that are not designed to initiate sessions with them.
- They are also known as "no login" accounts (nologin).
- They are created (usually) automatically at the time of installing Linux or the application.
- They are usually assigned a UID between 1 and 100 (defined in /etc/login.defs)

Normal users

- They are used for individual users.
- Each user has a working directory, usually located in / home.
- Each user can customize their work environment.
- They have only full privileges in their work directory or HOME.
- For security, it is always better to work as a normal user instead of the root user, and when you need to use commands only from root, use the su command.
- Current Linux distros are generally assigned a UID greater than 1000, starting with kernel 2.6.
- Each time a user is created, a group with the same name is created and the UID coincides with the GUID.

Configuration files

Normal and root users in their home directories have several files that start with "." That is, they are hidden. They vary a lot depending on the Linux distribution you have, but you will surely find the following or similar ones:

.bash_profile here we can indicate aliases, variables, configuration of the environment, etc. that we want to start at the beginning of the session.

.bash_logout here we can indicate actions, programs, scripts, etc., that we want to execute when leaving the session.

.bashrc is the same as .bash_profile, it is executed at the beginning of the session, traditionally in this file the programs or scripts to execute are indicated, unlike .bash_profile that configures the environment

STEP 1: Register users.

Add new users or create new accounts in Linux.

    useradd

a) Help.

    man useradd
    useradd --help

b) Register a user by default.
    The working directory is not established.
    You have to define the working directory with
    mkdir.
        ls
        mkdir /home/user4
        useradd user4
        cat /etc/passwd
        useradd user5
        passwd user5
        ctrl+alt+F3
        pwd --> /
        mkdir user5
c) Disconnect the user:
        logout
        exit
        login
    Register a user and create a directory at the same
    time.
c.1) Register and assign working directory. It should be
      created before the directory.
          useradd -d /PRÁCTICA/alumno alumno
          mkdir /PRÁCTICA
          mkdir /PRÁCTICA/alumno
c.2) Register and assign working directory. It should be
      created before the directory.
          useradd   -m -d /PRÁCTICA/alumno1
alumno1
d) Create a user by assigning a main group.
    -> I change the password to the user with whom I
    work.
        useradd -m –d /PRÁCTICA/alumno2   -g
    torero alumno2
        passwd alumno2
        passwd
e) Assign a user to different groups, be a secondary
    member of different groups.
        useradd -m -d /PRÁCTICA/alumno3 -g torero
–G bachillerato,eso alumno3.
        passwd alumno3
        groups alumno3
        gpasswd -A alumno3 eso
        groups alumno3
f) Establish the user identification -or number.
        useradd -m –d /PRÁCTICA/alumno4 -g eso -u 1500 alumno4
        passwd alumno4
        groups alumno4
        id alumno4
g) Assign the same identification to another user, it is the -o option.
        id root
        id
        useradd -m –d /PRÁCTICA/picador -g root -G torero –u 0 -o picador
        passwd picador
        id picador
h) Add a comment line to the file / etc / passwd.
        useradd -m –d /PRÁCTICA/alumno6 -g torero -c "alumno perezoso, sin ilusión" alumno6
        cat /etc/passwd
i) Assign the type of command interpreter (SHELL).
        useradd -m –d /PRÁCTICA/alumno7 -g torero -c "alumno perezoso, sin ilusión" -s /bin/sh alumno7
j) Establish restrictions on the assignment of the expiration date of the key.
        useradd –m –d /PRÁCTICAs/alumno8 -e 04/06/2014 alumno8
k) Set the number of days that a user's account will be enabled after the expiration or expiration date ("7 days").
        useradd -m -d /PRÁCTICAs/alumno9 -e 05/06/2014 –f 12 alumno9
    f 12 is the number of days, which can be accessed with the user after the key or account has expired.
l) Notification, prior to the expiration of an account -W number of days.
        useradd -m -d /PRÁCTICAs/alumno10 –e 05/07/2014 –f 12 –W 20 alumno10
    20 days before the expiration of the account, each time it is accessed it is reported that the account expired, in a period of time.
        useradd -m -d /PRÁCTICAs/alumno10 –e 05/07/2014 –f 12 –W 35 alumno10

useradd	
Syntax: useradd [options] USER_NAME	
OPTION	**DESCRIPTION**
-b	Base directory for the personal directory of the new account.
-c COMMENTARY	GECOS field of the new account.
-d DIR_PERSONAL	Personal directory of the new account.
-D	Print or change the default settings of useradd.
-e	Expiration date of the new account.
-f	Period of inactivity of the password of the new account
-g GROUP	Name or identifier of the primary group of the new account.
-G GROUPS	List of supplementary groups of the new account.
-k DIR_SKEL	Use this alternative skeleton directory.
-K	Overwrites the default values of /etc/login.defs.
-l	It does not add the user to the lastlog and faillog databases
-m	Create the user's personal directory.
-M	It does not create the user's personal directory.
-N	It does not create a group with the same name as the user.
-o	It allows creating users with duplicate (non-unique) identifiers (UIDs).
-p PASSWORD	Password encrypted for the new account.
-r	Create a system account.
-R CHROOT_DIR	Directory in which to do chroot.
-s CONSOLE	Access console of the new account.
-u UID	Identifier of the user of the new account.
-U	Create a group with the same name as the user.
-Z USER_SE	Use the user indicated for the SELinux user.

When creating a default user, a directory with the same name is assigned in the /home directory, which we should have previously created (mkdir). If the directory is not previously created, the user is assigned to a working directory that does not exist.
For the directory to be created at the same time, the options must be used:
        useradd -m -d /home/mialumno

> You can specify the key with -p, but you must give it encrypted. ---> use passwd

STEP 2: Modify a usermod user account.

All useradd options are used in the same way as it incorporates two options [-L | -U] allows you to block or unblock a user's account, just like with the passwd command, whose options are in lowercase.

      usermod

a)   Block and unblock a user account.

a.1) Block a user account.

      usermod -L alumno10

a.2) Unlock a user account.

      usermod -U alumno10

b)   Unlock a user account.

      login

      passwd

c)   Display active processes

      ps -aux

d)   Killing a process.

      kill -9 1926

> **NOTE:** If the user is connected, the process must be killed in order to expel the user.

STEP 3: Delete a user account.

Delete a user's account /etc/passwd, delete the user and all their data or only the user.

      userdel

a)   Help.

      userdel --help

b)   Delete only the user.

      userdel alumno10

c)   Delete all the information regarding the user of point b).

      /etc/passwd

      /etc/shadow

      /etc/group

      /PRÁCTICA/alumno10 **(delete the link), the information and the folder are not deleted.**

d)   Delete the user and all the content of their work.

      userdel -r alumno9

Delete the files + directories referenced by this user, as owner in their working directory (defect/home/student9).

userdel	
Syntax:	**userdel [options] user**
OPTION	**DESCRIPTION**
-a	Reports the status of passwords for all accounts.
-d	Clear the password for the indicated account -e -expire force to have the account password expire.
-f	Force the deletion of files, even if they do not belong to the user.
-h	Show this help message and it ends.
-k	Change the password only if it has expired.
-i	Sets the inactive password after expiring to INACTIVE.
-l	Sets the maximum number of days AS_MIN.
-q	Silent password mode a.
-r	Change the password in the REP repository.
-R	Directory in which to do chroot.
-S	Report the status of the password the account AS_MAX sets the maximum number AS_MAX before changing the password to indicated life
-Z	Remove any SELinux user mapping for the user.

STEP 4: Set the password to a user.

Request the password of a registered user, allow to block/unblock a user's account, they must be created.

      passwd

a)   Help.

      passwd --help

b)   Set the password to a user.

      passwd user4

c)   Block an account.

      passwd -l alumno8

d)   Unlock an account.

      passwd -u alumno8

e)   Borrar un password o clave.

      passwd -d alumno8

f)   Report the status of the password of all accounts.

      passwd -a

g)   Report the status of the password.

      passwd -S

      root P 05/20/2014 0 99999 7 -1

STEP 5: Identification of users and groups to which they belong.

Displays the user and group information for the specified USER NAME, or (when USER NAME is omitted) for the current user.

      id

a)   Help.

      id --help

b)   Default.

      id

They assume that what we ask is for the user with whom we work.

c) Ask for a specific user.
    id   alumno8
    id   root
    uid=0(root) gid=0(root) grupos=0(root)

d) Display the main group to which the user belongs.
    id -g alumno4

e) Visualize the secondary groups to which they belong.
    id -G alumno3

f) Identify the real permits.
    id -r
    id -r alumno3

g) Identify a user.
    id -u
    id -u alumno3

h) Display the user's name.
    id -n
    id -n -g
    id -n -G
    id -n -u
    id -n -g alumno3
    id -n -G alumno3
    id -n -u alumno3

id

Syntax:	id [options] user	
OPTION	**DESCRIPTION**	
-a	No effect, for compatibility with other versions.	
-Z	Displays only the security context of the current user.	
-g	Displays only the main group ID.	
-G	Shows only the supplementary, secondary groups.	
-n	Show a name instead of a number, for -ugG.	
-r	Shows the real ID instead of the effective ID, for -ugG.	
-u	Shows only the user's effective ID.	

chown

Syntax:	chown [option] new_user file_name / directory	
OPTION	**DESCRIPTION**	
-R	Change the permission in files that are in subdirectories of the directory in which you are at that moment.	
-c	Change the permission for each file	
-f	Prevents chown from displaying error messages when it is unable to change the ownership of a file.	

STEP 6: Identify the name of the group (s) to which a user belongs.

It shows the belonging to the group of each USER, or, if the user's name is not specified, the current process (which may be different if the group database has changed).
    groups

a) Help.
    groups --help

b) Default.
    groups alumno
    alumno :   *alumno adm cdrom sudo dip plugdev lpadmin sambashare*

groups

Syntax: groups usuario

STEP 7: Show the name of the current user.

    logname

a) Help.
    logname --help

b) Default.
    logname

c) Order version.
    logname --version

logname

Displays the name of the current user.

STEP 8: Establish or change the owner of a file or directory.

It is used to change the owner /user of the file or directory. It is an administrator command, only the root user can change the owner of a file or directory.
    chown

a) Help.
    chown   --help

b) Set the user and group only to the root user for the /backup directory:
    chown root:root /backup

c) The owner of the file "texto.txt" is root, it changes to the new user baldo.
    chown baldo texto.txt

d) The owner of the "texto01.txt" directory is root, with the -R option the user of the files and subdirectories is also changed.
    chown -R baldo   texto01.txt

e) Here the owner changes only for the file "texto02.txt".
    chown -c   baldo texto02.txt

f) Set the root user as owner and allow any member of the ftp group that has access to the thing.txt file (verify that you have sufficient write/read permissions).
    chown root:ftp /home/data/cosa.txt

g) Set the owner to the name of any user and a group.
    chown root:ftp /home/data/cosa.txt

h) Set the owner to the name of a user in any group.
    chown root:ftp /home/data/cosa.txt

i) Set the owner to be no user of any group.
    chown root:ftp /home/data/cosa.txt

SUID It allows to execute a file, and it is executed as if the one executing it was the owner of the File.
        chmod o+s   fichero

STEP 9: Verify the integrity of the password files.

      pwck

a)   Help.

      pwck --help

b)   Check the integrity without options.

      pwck /etc/passwd

      user 'lp': directory '/var/spool/lpd' does not exist

      user 'news': directory '/var/spool/news' does not exist

      user 'uucp': directory '/var/spool/uucp' does not exist

      user 'www-data': directory '/var/www' does not exist

      user 'list': directory '/var/list' does not exist

      user 'irc': directory '/var/run/ircd' does not exist

      user 'gnats': directory '/var/lib/gnats' does not exist

      user 'nobody': directory '/nonexistent' does not exist

      user 'syslog': directory '/home/syslog' does not exist

      user 'whoopsie': directory '/nonexistent' does not exist

c)   Display the output only with the error reports.

      pwck -q /etc/passwd

d)   Execute in read only mode.

      pwck -r /etc/passwd

e)   Sort the entries by UID in the file passwd and shadow.

      pwck -s /etc/passwd

pwck

Syntax: pwck [Opción] [passwd [shadow]]

OPTION	DESCRIPTION
-q	Report only the Errors.
-r	Run the pwck command in read-only mode.
-s	Sort the entries in **/etc/passwd** and **/etc/shadow** by UID.

STEP 10: Verify the integrity of the group's files.

      grpck

a)   Help.

      grpck --help

b)   Information by default.

      grpck

c)   Execute only in reading mode.

      grpck -r /etc/group

d)   Execute the entries in /etc/group and sort them by GID and compare them with gshadow.

      grpck -s /etc/group

grpck

Syntax: grpck [-r] [-s] [group [gshadow]]

OPTION	DESCRIPTION
-r	Run in read-only mode. This makes all questions regarding the changes that need to be answered without any user intervention.
-s	Type text or the address of a website, or translate a document. Sort the entries in /etc/group and /etc/gshadow by GID.

PRÁCTICA 24: Administrar usuarios y grupos en Linux con scripts Perl.
DESCRIPCIÓN:

Perl was created by Larry Wall to simplify administration tasks of a Unix system, although nowadays it has become one of the best tools for scripting or building websites.

Perl is a fast language despite being interpreted, multiplatform and has a large number of libraries for the development of almost any type of application. It's free software, you do not have to pay it

The simplest example of a program in Perl, would be in Linux:

```
#!/bin/perl
print "Hola Mundo\n";
```

Now, we take the previous program and save it in a file with extension .pl that is the extension of the scripts in Perl. To execute it, simply call it by its name (if you have the appropriate execution permissions):

```
./hola.pl
```

As it can be deduced, with the first line we indicate where is the Perl interpreter, after #! This would not be necessary if, when executing the script, we do it using the interpreter. For example:

```
/usr/bin/perl hola.pl
```

The adduser and addgroup commands add users to the system according to the command line options and the configuration in /etc/adduser.conf. They provide a more user-friendly interface for useradd and groupadd programs, choose values for UID and GID compliant with Linux standards, create a personal directory with the default settings.

> **THERE IS A DIFFERENCE BETWEEN EXECUTING A SCRIPT BASH USING ./ or SH**
> a) When you execute a script by passing the file name of the script to an interpreter (sh, python, perl, etc.), you are actually executing the interpreter by passing the program as an argument, all automatically and without the user that executed it the script finds out.
> ```
> # sh  /home/user1/prueba003.sh
> ```
> b) To execute a script on its own, 2 conditions must be met:
> 1. The script must include a "bang line". This is the first line of a script, which should start with the characters #! and that you must specify the path in which the interpreter is located. It is important to note that this condition is true for any type of script (python, perl, etc.), not just those of bash.
> ```
> #!/bin/bash
> #!/bin/perl
> ```
> 2. The file must have execute permissions.
> ```
> chmod 744  prueba003.sh
> ./prueba003.sh
> . prueba003.sh
> ```

STEP 1: Register a group.

The script that allows you to create a group in the command line.

```
addgroup
```

a) Help.

```
addgroup  --help
```

b) By default it gives us an error, missing parameters

```
root@profesor:/home/alumno# addgroup
addgroup: Sólo se permiten uno o dos nombres.
```

c) Create a group by only establishing the name, it is assigned by default the gid of the group according to the counter.

```
root@profesor:/home/alumno# addgroup nuevo
Añadiendo el grupo 'nuevo' (GID 1006) ...
Hecho.
```

d) Create a group by setting the group's gid and name.

```
root@profesor:/home/alumno# addgroup—gid 1400  fiestas
Añadiendo el grupo 'fiestas' (GID 1400) ...
Hecho.
```

STEP 2: Register a normal user in UBUNTU.

Add a system user.

```
adduser
```

> **Register a normal user**
> **adduser** [--home DIR] [--shell SHELL] [--no-create-home] [--uid ID] [--firstuid ID] [--lastuid ID] [--gecos GECOS] [--ingroup GROUP | --gid ID] [--disabled-password] [--disabled-login] [--encrypt-home] USER

a) Help.

```
adduser --help
```

b) Default.

```
root@profesor:/home/alumno# adduser
adduser: Sólo se permiten uno o dos nombres.
root@profesor:/home/alumno# adduser alumno00
Añadiendo el usuario 'alumno00' ...
Añadiendo el nuevo grupo 'alumno00' (1007) ...
Añadiendo el nuevo usuario 'alumno00' (1004) con grupo 'alumno00' ...
Creando el directorio personal '/home/alumno00' ...
Copiando los ficheros desde '/etc/skel' ...
Introduzca la nueva contraseña de UNIX:
Vuelva a escribir la nueva contraseña de UNIX:
passwd: contraseña actualizada correctamente
Cambiando la información de usuario para alumno00
Introduzca el nuevo valor, o presione INTRO para el predeterminado
Nombre completo []: primer alumno
Número de habitación []: 00
Teléfono del trabajo []: 034910000001
Teléfono de casa []: 034910000002
Otro []:
¿Es correcta la información? [S/n] S
```

STEP 3: Register a user of the system in UBUNTU.

a) Add one more user by default to the system.

```
root@profesor:/home/alumno# adduser-system alumno12
Añadiendo el usuario del sistema 'alumno12' (UID 121) ...
Añadiendo un nuevo usuario 'alumno12' (UID 121) con grupo 'nogroup' ...
Creando el directorio personal '/home/alumno12' ...
```

b) Add the username and the directory associated with it.

```
adduser   --system --home /home/alumno02 alumno02
```

c) Add a user and his numerical identification.

```
adduser   --system --home /home/alumno03 alumno03
```

d) Add a user and their numerical identification.

```
adduser   --system --no-create-home --uid 1200 alumno030
```

e) Add a user and his numerical identification.

```
adduser   --system --home /home/alumno04 --uid 1250 alumno04
```

f) Add a user and their numerical identification and group identification.

```
adduser   --system --home /home/alumno05   --ingroup fiestas alumno05
adduser   --system --home /home/alumno06   --gid 1400 alumno06
adduser   --system --home /home/alumno07   --ingroup users alumno07
Añadiendo el usuario del sistema 'alumno07' (UID 123) ...
Añadiendo un nuevo usuario 'alumno07' (UID 123) con grupo 'users' ...
Creando el directorio personal '/home/alumno07' ...
adduser   --system --home /home/alumno08   --gid 1400 alumno08
Añadiendo el usuario del sistema 'alumno08' (UID 122) ...
Añadiendo un nuevo usuario 'alumno08' (UID 122) con grupo 'fiestas' ...
Creando el directorio personal '/home/alumno08' ...
```

g) Add a user and their numerical identification, main work group, main group and Shell.

```
adduser   --system --home /home/alumno09   --shell /bin/csh --uid 1410 --ingroup
fiestas  alumno09
Añadiendo el usuario del sistema 'alumno09' (UID 1410) ...
Añadiendo un nuevo usuario 'alumno09' (UID 1410) con grupo 'fiestas' ...
Creando el directorio personal '/home/alumno09' ...
```

> **Add a system user**
> adduser --system [--home DIR] [--shell SHELL] [--no-create-home] [--uid ID] [--gecos GECOS] [--group | --ingroup GROUP | --gid ID] [--disabled-password]
> [--disabled-login] USER

h) Add a user and his numerical identification, and deactivate the password.

```
adduser   --system --home /home/alumno10 --uid 1300 --disabled-password alumno10
Añadiendo el usuario del sistema 'alumno10' (UID 1300) ...
Añadiendo un nuevo usuario 'alumno10' (UID 1300) con grupo 'nogroup' ...
Creando el directorio personal '/home/alumno10' ...
```

i) Add a user and deactivate the account.

```
adduser   --system --home /home/alumno11 --disabled-password --disabled-login alumno11
Añadiendo el usuario del sistema 'alumno11' (UID 124)   ...
Añadiendo un nuevo usuario 'alumno11' (UID 124) con grupo 'nogroup' ...
Creando el directorio personal '/home/alumno11' ...
```

STEP 4: Register a normal user in Slackware 14.1

```
root@192:~# adduser
Login name for new user []: alumno020
User ID ('UID') [ defaults to next available ]: 1800
Initial group [ users ]:
Additional UNIX groups:
Users can belong to additional UNIX groups on the system.
For local users using graphical desktop login managers such
as XDM/KDM, users may need to be members of additional groups
to access the full functionality of removable media devices.

* Security implications *
Please be aware that by adding users to additional groups may
potentially give access to the removable media of other users.
If you are creating a new user for remote shell access only,
users do not need to belong to any additional groups as standard,
so you may press ENTER at the next prompt.

Press ENTER to continue without adding any additional groups
Or press the UP arrow key to add/select/edit additional groups
:

Home directory [ /home/alumno020 ]
Shell [ /bin/bash ]
Expiry date (YYYY-MM-DD) []:
New account will be created as follows:

Login name.......:    alumno020
UID..............:    1800
Initial group....:    users
Additional groups:    [ None ]
Home directory...:    /home/alumno020
Shell............:    /bin/bash
Expiry date......:    [ Never ]
This is it... if you want to bail out, hit Control-C.  Otherwise, press
ENTER to go ahead and make the account.
```

```
Creating new account...

Changing the user information for alumno020
Enter the new value, or press ENTER for the default
Full Name []: alumno creado en Slackware 14.1
Room Number []: 1000
Work Phone []: 910000001
Home Phone []: 910000002
Other []:
Changing password for alumno020
Enter the new password (minimum of 5 characters)
Please use a combination of upper and lower case letters and numbers.
New password:
Re-enter new password:
passwd: password changed.

Account setup complete.
```

STEP 5: Register a normal user in CentOS 7.0 with adduser.

a) Create a user, and assign a working directory.
   [root@localhost home]# adduser -m -b /home alumno003
   [root@localhost home]# ls -l
   total 0
   drwx------. 2 alumno   alumno   59 ago 6 01:34 alumno
   drwx------. 2 alumno003 alumno003 59 sep 13 18:41 alumno003

b) Does not add the user to the lastlog and faillog database.
   [root@localhost home]# adduser -l -m -b /home alumno004
   [root@localhost home]# ls -l
   total 0
   drwx------. 2 alumno   alumno   59 ago 6 01:34 alumno
   drwx------. 2 alumno003 alumno003 59 sep 13 18:41 alumno003
   drwx------. 2 alumno004 alumno004 59 sep 13 18:44 alumno004

c) Do not create a group with the same name as the user.
   [root@localhost home]# adduser -N -m -b /home alumno005
   [root@localhost home]# ls -l
   total 0
   drwx------. 2 alumno   alumno   59 ago 6 01:34 alumno
   drwx------. 2 alumno003 alumno003 59 sep 13 18:41 alumno003
   drwx------. 2 alumno004 alumno004 59 sep 13 18:44 alumno004
   drwx------. 2 alumno005 users   59 sep 13 18:46 alumno005

d) Create a system account.
   [root@localhost home]# adduser -r -m -b /home alumno006
   [root@localhost home]# ls -l
   total 0
   drwx------. 2 alumno   alumno   59 ago 6 01:34 alumno
   drwx------. 2 alumno003 alumno003 59 sep 13 18:41 alumno003
   drwx------. 2 alumno004 alumno004 59 sep 13 18:44 alumno004
   drwx------. 2 alumno005 users   59 sep 13 18:46 alumno005
   drwx------. 2 alumno006 alumno006 59 sep 13 18:47 alumno006

e) Create a user with the same ID or user ID as others.
   [root@localhost home]# adduser -u 0 -o -m -b /home alumno007
   [root@localhost home]# ls -l
   total 0
   drwx------. 2 alumno   alumno   59 ago 6 01:34 alumno
   drwx------. 2 alumno003 alumno003 59 sep 13 18:41 alumno003
   drwx------. 2 alumno004 alumno004 59 sep 13 18:44 alumno004
   drwx------. 2 alumno005 users   59 sep 13 18:46 alumno005
   drwx------. 2 alumno006 alumno006 59 sep 13 18:47 alumno006
   drwx------. 2 root   alumno007 59 sep 13 18:50 alumno007

f) Create a new connection shell.
   adduser -s /bin/sh -m -b /home alumno008

g) Establish the main group.
   cat /etc/group
   adduser -g root -m -b /home alumno009

h) Set the secondary groups and the main group and the directory to create the working directory.
   adduser -g users -G alumno001,alumno002,alumno003 -m -b /home alumno010

i) Establish a comment line.
   adduser -g users -G alumno001,alumno002,alumno003 -c 'alumno 11 nuevo' -m -b /home alumno011
   cat /etc/passwd

```
...
alumno:x:1000:1000:alumno:/home/alumno:/bin/bash
alumno001:x:1001:1001::/home/alumno001/alumno001:/bin/bash
alumno002:x:1002:1002::/home/alumno002/alumno002:/bin/bash
alumno003:x:1003:1003::/home/alumno003:/bin/bash
alumno004:x:1004:1004::/home/alumno004:/bin/bash
alumno005:x:1005:100::/home/alumno005:/bin/bash
```

---

adduser [opciones] USER
adduser -D
adduser -D [options]

OPTION	DESCRIPTION
-b	Base directory for the personal directory of the new account.
-c	GECOS field of the new account.
-d	Personal directory of the new account.
-D	Print or change the default settings of useradd.
-e	Expiration date of the new account.
-f	Period of inactivity of the password of the new account.
-g	Name or identifier of the primary group of the new account.
-G	List of supplementary groups of the new account.
-k	Overwrites the default values of "/etc/login.defs".
-K	It does not add the user to the lastlog and faillog databases.
-l	It does not add the user to the lastlog and faillog databases.
-m	Create the user's personal directory.
-M	It does not create the user's personal directory.
-N	It does not create a group with the same name as the user.
-o	It allows creating users with duplicate (non-unique) identifiers (UIDs).
-p	Password encrypted for the new account.
-r	Create a system account.
-R	Enter the chroot directory.
-s	Access console of the new account.
-u	Identifier of the user of the new account.
-U	Create a group with the same name as the user.
-Z	Use the user indicated for the SELinux user.

NOTE: the path to create the working directory only writes the parent directory /home, (specifies -m to create the directory -b is the path of the parent directory), from it, creates the directory studentXXX, which is the last written value. But the parent directory path is assigned.
   adduser alumno013
It is created by default in the /home directory
   /home/alumno013.

```
alumno006:x:998:997::/home/alumno006:/bin/bash
alumno007:x:0:1005::/home/alumno007:/bin/bash
alumno008:x:1006:1006::/home/alumno008:/bin/sh
alumno009:x:1007:0::/home/alumno009:/bin/bash
alumno010:x:1008:100::/home/alumno010:/bin/bash
alumno011:x:1009:100:alumno 11 nuevo:/home/alumno011:/bin/bash
```

j) Add a user with restrictions of expiration date and inactivation date.
```
adduser  -c 'alumno 12 con restricciona acceso 60 dias caducidad 15' -e 13/11/2015 -f 15 -m -b
/home  alumno012
```

k) Check if there are restrictions.
```
cat /etc/shadow

...

alumno:$6$XI3WRSkTGwx6Z1.D$HazdTO2MJqMrbvgt01N1LSTYkXUN0XQ../6j2vJvmVh7lkrtnwzPE74VvwMTh5vThToNPn
CVCwCd1uD0mDwOB/:16652:0:99999:7:::
alumno001:!!:16691:0:99999:7:::
alumno002:!!:16691:0:99999:7:::
alumno003:!!:16691:0:99999:7:::
alumno004:!!:16691:0:99999:7:::
alumno005:!!:16691:0:99999:7:::
alumno006:!!:16691::::::
alumno007:!!:16691:0:99999:7:::
alumno008:!!:16691:0:99999:7:::
alumno009:!!:16691:0:99999:7:::
alumno010:!!:16691:0:99999:7:::
alumno011:!!:16691:0:99999:7:::
alumno012:!!:16691:0:99999:7:15:16811:
```

l) Assign passwords, to users already created students.
```
passwd  alumno001

...

passwd  alumno012
cat /etc/shadown

...

alumno010:$6$0UMfiYjt$a2tW5f0avkh7gaEJsoPnIm3jzAyQZ5cEgzvLKGRoca9/WaeiE7U9Qwi7dzbiScrBCzrnYlVnJRq
7b3NRWcxsX/:16691:0:99999:7:::
alumno011:$6$Ru0AEBpG$pTQkt0rUfs3kmS6GCC4krgjSl8rVwsjgk9ifCFiK0Y22pdHpqs28OSnSdLBlSkr1.uMqsdQsyuA
Sx6bgTS/9E/:16691:0:99999:7:::
alumno012:$6$yx3XY0e3$r.wFDcUYL70DgAVUoZegdBAvjvrKoE/pXV7ASJ5N5l/ltO.8ujMITdCmF6xo.WaEnEXd.i1Y8FL
ihbqo7Y26o0:16691:0:99999:7:15:16811:
```

WORK UNIT VI: System Administration II. System settings

PRACTICE 25: Device information in Linux.

PRACTICE 26: Processes and operations.

Commands

free, df, du, file,
vmstat, pmap, ps,
sar, time lock,
pstree, top, kill,
sleep, bg, fg, jobs,
nice, renice, nohup,
stop, init, telinit,
service.

Content:
- Disk management in Linux.
- Memory management in Linux.
- Update of the operating system.
- Manage operating system hardware.
- Monitoring and system performance.
- Add/Remove/Update software in the operating system.
- Task programming in Linux.

PRACTICE 25: Device information in Linux

DESCRIPTION:

Memory.
Disk space
Shared Virtual Memory

There is a need to share memory between processes. There may be several processes running in the system, processes such as the bash shell command, rather than multiple copies of bash, each with its own virtual memory address space, it would certainly be much better "to have only one copy in memory and that all the processes that run bash share it. " Swap memory space or Swap is what is known as virtual memory. The difference between real and virtual memory is that the latter uses space in the storage unit instead of a memory module. When the actual memory runs out, the system copies part of its contents directly into this exchange memory space in order to perform other tasks.

Using virtual memory has the advantage of providing the necessary additional memory when the actual memory has been exhausted and a process has to be continued. As a consequence of using space in the storage unit as memory is that it is considerably slower.

STEP 1: Memory layout.

The free command shows information about the free and used memory of the system.

    free

a) Help.

    free --help

b) Default.

    free

	total	usado	libre	compart.	búffers	almac.
Mem:	1025940	868832	157108	2272	99188	373444
-/+ buffers/cache:		396200	629740			
Intercambio:	1046524	440	1046084			

c) Visualize the output in different unit formats.

    free -b
    free -k
    free -m
    free -g
    free --tera

d) Visualize with details.

    free -l

	total	usado	libre	compart.	búffers	almac.
Mem:	1025940	868836	157104	2272	99188	373444
Bajo:	890828	762136	128692			
Alto:	135112	106700	28412			
-/+ buffers/cache:		396204	629736			
Intercambio:	1046524	440	1046084			

e) Visualize the entire memory.

    free -t

	total	usado	libre	compart.	búffers	almac.
Mem:	1025940	868832	157108	2272	99188	373444
-/+ buffers/cache:		396200	629740			
Intercambio:	1046524	440	1046084			
Total:	2072464	869612	1202852			

f) Use old format.

	total	usado	libre	compart.	búffers	almac.
Mem:	1025940	868832	157108	2272	99188	373444
Intercambio:	1046524	440	1046084			

STEP 2: File system space.

    df

a) Help.

    df --help

b) Default.

    df

```
Filesystem        1K-blocks    Used Available Use% Mounted on
/dev/sda1         24639868 8349448  15015748  36% /
tmpfs               506240       0    506240   0% /dev/shm
```

c) Include file systems that have blocks 0.

```
root@192:~# df -a
Filesystem        1K-blocks    Used Available Use% Mounted on
/dev/sda1         24639868 8349448
/proc
sysfs                    0       0         0    - /15015748  36% /
proc                     0       0         0    - sys
tmpfs               506240       0    506240   0% /dev/shm
```

STEP 3: Space used by the files.

    du

a) Help.

    du --help

free

Syntax:	free [options]
OPTION	**DESCRIPTION**
-b	Show the output in bytes.
-k	Show the output in kilobytes.
-m	Show the output in megabytes.
-g	Show the output in gigabytes.
--tera	Show the output in terabytes.
-h	It shows output in human readable format.
--si	Repeat the exit N times and then finish.
-l	Repeat the output every N seconds.
-o	Show the total for RAM + swap.
-t	Use old format (without line - / + buffers / cache).
-s N	Show detailed statistics of low and high memory.
-c N	Use powers of 1000 not 1024.

df

Syntax:	df [option]... [file]...	
OPTION	**DESCRIPTION**	
-a	Include file systems that have blocks 0	
-B	SIZE size blocks bytes	
-h	Printing sizes in human readable format (for example, 1K 234m 2G)	
-H	Similarly, but the use of powers of 1000 not 1024	
-i	List inode information instead of block use	
-k	As size --block = 1K	
-l	Limit list to local file systems.	
--no-sync	Do not invoke synchronization before obtaining usage information (by default).	
-P	Use the POSIX output format.	
-t	Limit list to TYPE file systems.	
-T	Display the type of file system.	
-x	Limit the list of file systems not of type TYPE.	

b) Default.

    du

c) Display all the information.

    du -a

d) Display all the information of a directory.

    du -a   alumnos

e) Display only the total space used.

    du -s

f) Shows the size of each file in the specified directory.

    du -s   curso2014

g) Shows the total disk space used by the specified directory.

    du -h

h) Shows the capacity of the current folder.

    du -h practica.doc

du	
Syntax:	**du [option]... [file]**
OPTION	**DESCRIPTION**
-a	Write the count of all the files, not just the directories.
-B	Use SIZE size blocks bytes.
-b	Size of the print in bytes.
-c	Produce a great total.
-D	Files that are reference for symbolic links.
-h	Print sizes in human readable format (for example, 1K 234m 2G).
-H	The use of the unit of measure is 1000 not 1024
-k	The size --block = 1K
-l	Count the sizes many times if hard linked.
-L	Remove the reference of all the symbolic links.
-S	They do not include the size of subdirectories.
-s	Show only a total for each argument.

STEP 4: Infomate if the object you see is a directory or a file.

The file command tells you if the object you see is a directory or a file.

    file

a) Help.

    file --help

b) Default.

    file /etc/passwd

    file /etc/init.d/networking

c) View information in a directory.

    file *

Descargas: directory
Documentos: directory Literales
Escritorio: directory
examples.desktop: UTF-8 Unicode text

file	
Syntax:	**file [options] filename / directory**
OPTION	**DESCRIPTION**
-c	Check the magic file for formatting errors. For reasons of efficiency, this validation is usually not carried out.
-h	It does not follow symbolic links.
-m	Use mfile as an alternative magic file.

STEP 5: Shows the status of the virtual memory (swap partition).

    vmstat

a) Help.

    vmstat --help

b) Visualize the virtual memory by default.

    vmstat

c) If the argument is a number, it specifies the interval in seconds for the list to be repeated.

    vmstat 5

Displays the information every five seconds.

STEP 6: Information on the use of memory.

The file located in the /proc/meminfo directory contains all the information about the use of your memory.

    meminfo

a) List the contents of the /proc/meminfo file.

    cat   /proc/meminfo

    less /proc/meninfo

STEP 7: Show or examine the memory map and the libraries of a process.

On a Linux server, you can easily list the details of an active process and view its actual memory consumption. Sometimes we see that the computer slows down and we have to be able to know which process is saturating the RAM. Once logged in as root.

    pmap pmap

a) Help.

    pmap --help

b) Request the memory map of a process (PID).

    ps -aux

    pmap   2281

c) Request the memory map of a process and the details.

    pmap   2281 -x

pmap	
Syntax:	**pmap [-rsIF] [pid	core] ...** **pmap -x [-aslF] [pid	core] ..**
OPTION	**DESCRIPTION**
-a	Print anonymous and exchange reservations to share assignments.
-F	It grabs the target process, even if another process has control.
-l	Displays unresolved dynamic linker map names.
r	Print reserved addresses of the process.
-s	Print HAT page size information
-S	They exchange reservation information for the cartography.
-x	Additional information by cartography
Location directory /usr/bin.	

2281: su

Dirección	Kbytes	RSS	Sucio	Modo	Asignaciones
0008048000	32	28	0	r-x--	su
0008050000	4	4	4	r----	su
0008051000	4	4	4	rw---	su

0008052000	*16*	*12*	*12*	*rw---*	*[anon]*
00082cc000	*132*	*76*	*76*	*rw---*	*[anon]*
00b70d0000	*96*	*56*	*0*	*r-x--*	*libpthread-2.19.so*
00b70e8000	*4*	*4*	*4*	*r----*	*libpthread-2.19.so*
00b70e9000	*4*	*4*	*4*	*rw---*	*libpthread-2.19.so*

d) Request the memory map of a process and all the details.

    pmap   2281 -X

    2281: su

Dirección	Perm	Desplazamiento	Dispositivo	Inodo	Size	Rss	Pss	Referenced	Anonymous	Swap	Locked	Asignaciones
08048000	r-xp	00000000	08:01	131230	32	28	28	28	0	0	0	su
08050000	r—p	00007000	08:01	131230	4	4	4	0	4	0	0	su
08051000	rw-p	00008000	08:01	131230	4	4	4	4	4	0	0	su
08052000	rw-p	00000000	00:00	0	16	12	12	0	12	0	0	
082cc000	rw-p	00000000	00:00	0	132	76	76	24	76	0	0	[heap]
b70d0000	r-xp	00000000	08:01	918613	96	56	0	56	0	0	0	libpthread-2.19.so
b70e8000	r—p	00017000	08:01	918613	4	4	4	0	4	0	0	libpthread-2.19.so

STEP 8: Display paging statistics.

        sar

a)   Help.

        sar

b)   Display the paging statistics.

        sar -B

> Install the packages: sysstat, atsar
>   **apt-get   install   sysstat**
>   **apt-get   install   atsar**
> Previous update
>   **apt-get   upgrade**

STEP 9: Display information on the system page.

        time

a)   Help.

        time   --help

b)   Information on the default page.

        time
        /usr/bin/time

> You must specify the fully qualified path of the "/usr/bin/time" command to avoid using the "time" command of the bash shell.

c)   Shows the page size of the system, page errors, etc. of a process during its execution.

        time -v
        /usr/bin/time -v

STEP 10: Free page information.

The file located in / proc / freepages contains information of the "free pages" of the virtual memory.

        cat /proc/sys/vm/freepages

It is possible to increase/decrease this limit: echo 300 400 500> /proc /sys/vm/ freepages.

STEP 11: Block a terminal.

It allows to block the terminal, for it asks for a password, twice.

        lock

a)   Help.

        lock --help

b)   Block the terminal by default.

        lock

PRACTICE 26: Processes and operations.
DESCRIPCIÓN:

a) Linux processes.

a.2) **Operations with processes.**
- See processes.
- Kill process
- Execute processes in the foreground.
- Run processes in the background.
- See the list of processes in the background.

b) Process management in Android.

b.1) Introduction.

Android OS is an operating system developed by Google for use on mobile devices. This means that it has been designed for systems with low memory and a processor that is not as fast as desktop processors. Android is based on the Linux 2.6 kernel. There are important modifications that have been made in the core, but it has the same core. The Android operating system is designed as a single user of the operating system, so Android takes advantage of this and runs each component as a different user. This allows Android to use the Linux security model and keep the processes in their own sandbox.

b.2) Description of the processes.

Process management in a typical operating system involves many data structures and complex algorithms, but does not go much beyond the level of the typical data structure process management. Android is similar in that at the base level the control structures look the same.

b.3) Process structure.

This data structure is managed by a standard process management, which is something like this:

Android OS ends a process when there is not enough memory for other processes.

All application components that are running in the process that is being terminated by the operating system are destroyed.

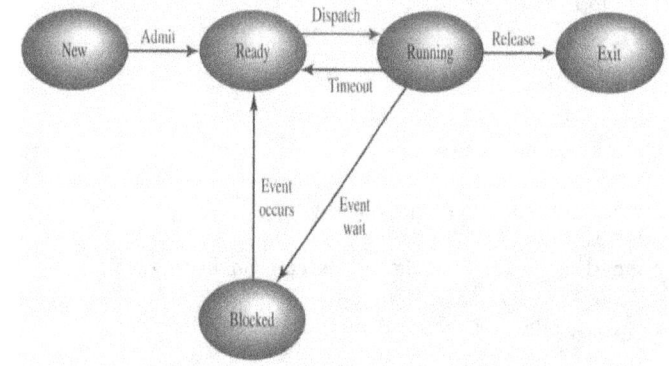

A new process will be initiated by those components when these components should work again Android OS decides which process to finish based on its relative importance to the user, for example, all components of a process are not visible.

When an application component is started and the application does not have any other component running, the Android system starts a new Linux process for the application with a single thread of execution. By default, all components of the same application are executed in the same process and thread (called the "main" thread). If an application component is started and a process already exists for that application (because another component of the application exists), then the component starts within that process and uses the same thread of execution. However, you can make arrangements for different components of the application to run in separate processes, and you can create additional threads for any process.

b.4) Life cycle processes.

The Android system tries to maintain an application process for as long as possible, but with the necessary time to eliminate the old processes to reclaim memory for new or more important processes. To determine which processes to follow and which to kill, the system places each process in a "hierarchy of importance", based on the components that are executed in the process and the state of the components. Minor processes are eliminated first, then those with the next lowest importance, and so on, as necessary to recover system resources.

STEP 1: Hierarchy of processes.

        ps
a)   Visualize processes ps.
a.1) Help.
        ps --help
a.2) Display by default.
        ps
a.3) Complete visualization of all processes.
        ps -aux
b)   Visualize the pstree process tree.
b.1) Structure of the process tree.
        pstree --help
b.2) Visualize defects in the tree structure.
        pstree
b.3) Display with arguments.
        pstree -a

ps [options]	
-a	List information about all the most frequently requested processes: all except process group leaders and processes not associated with a terminal.
-A ó e	List information for all processes.
-d	Lists information about all processes except session leaders.
-e	List information about all running processes.
-f	Generates a complete list.
-j	Show session identifier and process group.
-l	Generates a long list.

b.4) Visualize the processes of a user.

  pstree -u root

b.5) Visualize the processes with your PID.

  pstree -p

b.6) Visualize from a specific process.

  pstree PID

  pstree 0

  pstree 438

b.7) Sample organized by PID.

  pstree -n

  pstree -n -p

b.8) It shows a process by PID.

  pstree 1001

```
├─gnome-terminal(1838)─┬─gnome-pty-helpe(1839)
│                      │ ─bash(1840)───su(1863)───bash(1872)
─────pstree(19+
                       │         └─{gnome-terminal}(1841)
```

c) Visualize the processes with a top application.

  top

  ctop   → Use for graphic environment

  atop

  mtop

> A **process** is not a program but a running program, the program's stack, the variables that change their value, etc.
> A **daemon** of a multiuser system is the one that is always running in the background, since the system is started until it shuts down, so it is said that they are alive. Some demons are:
> - **Cron**. It is responsible for executing the programs that we indicate with the crontab command at certain times.
> - **Sendmail**. Employee to manage the email.
> - **Inetd**. It is the Internet super-demon.

> **Columns of information order ps.**
> **PID:** process identification number.
> **PGID:** Identification of the group or groups to which the father belongs.
> **SID:** Identification of the session (session process number).
> **TTY:** Type of device.
>   pts/0 -> input/output device.
> **TIME:** running time.
> **CMD:** order in execution.

STEP 2: Attributes of a process.

  ps

a) Active processes in the terminal.

  ps

b) It shows all the processes that are executed in the terminal.

  ps -a

c) Visualize all the processes that are running in the terminal.

  ps -A

  ps -e

d) View information with task control format.

```
ps  -j
PID     PGID    SID     TTY         TIME      CMD
1863    1863    1840    pts/0    00:00:00     su
1872    1872    1840    pts/0    00:00:00     bash
1950    1950    1840    pts/0    00:00:00     ps
```

e) Identification related to the processes that are being executed.

```
ps -r
PID     TTY         STAT    TIME    COMMAND
1952    pts/0       R+      0:00    ps -r
```

f) Visualize all processes that are running for an owner or user.

  ps -u 0

  ps U root

g) Most useful sequence.

  ps aux

> **STAT: State**
> **R:** running processes.
> **d:** sleep process (uninterrupted sleep).
> **n:** low priority process.
> **s:** asleep or waiting.
> **t:** process followed or stopped.
> **z:** zombie
> **w:** exchange process.
> **x:** (dead) Processes stopped (DEAD).
>
> **Execution plan, priority**
> **+** Execution in the foreground.
> **<** High priority.
> **N** No low priority
> **I** There are multiple threads.
> **L** blocked in Memory (REAL TIME I / O).
> **r** resident in memory.

h) See all the processes with all identifications also a column named STAT appears, with the execution status.

```
ps j U root |more
```

PPID	PID	PGID	SID	TTY	TPGID	STAT	UID	TIME	COMMAND
0	1	1	1	?	-1	Ss	0	0:00	/sbin/init
0	2	0	0	?	-1	S<	0	0:00	[kthreadd]
2	3	0	0	?	-1	S<	0	0:00	[migration/0]
2	4	0	0	?	-1	S<	0	0:00	[ksoftirqd/0]
2	5	0	0	?	-1	S<	0	0:00	[watchdog/0]
2	6	0	0	?	-1	S<	0	0:00	[events/0]
2	7	0	0	?	-1	S<	0	0:00	[cpuset]
2	8	0	0	?	-1	S<	0	0:00	[khelper]
2	9	0	0	?	-1	S<	0	0:00	[netns]
2	10	0	0	?	-1	S<	0	0:00	[async/mgr]
2	11	0	0	?	-1	S<	0	0:00	[kintegrityd/0]
2	12	0	0	?	-1	S<	0	0:00	[kblockd/0]
2	13	0	0	?	-1	S<	0	0:00	[kacpid]
2	14	0	0	?	-1	S<	0	0:00	[kacpi_notify]
2	15	0	0	?	-1	S<	0	0:00	[kacpi_hotplug]
2	16	0	0	?	-1	S<	0	0:00	[ata/0]
2	17	0	0	?	-1	S<	0	0:00	[ata_aux]
2	18	0	0	?	-1	S<	0	0:00	[ksuspend_usbd]
2	19	0	0	?	-1	S<	0	0:00	[khubd]
2	20	0	0	?	-1	S<	0	0:00	[kseriod]
2	21	0	0	?	-1	S<	0	0:00	[kmmcd]

> **PPID:** Identification of the parent process.
> **SID:** Session identification.
> **UID:** user identification.
> **TPGID:** id. of group of task.

2	22	0	0 ?	-1		S<	0	0:00	[bluetooth]	

ps -aux

USER	PID	%CPU	%MEM	VSZ	RSS	TTY	STAT	START	TIME	COMMAND
root	1	0.0	0.2	4460	2540	?	Ss	sep10	0:01	/sbin/init
root	2	0.0	0.0	0	0	?	S	sep10	0:00	[kthreadd]
root	3	0.0	0.0	0	0	?	S	sep10	0:06	[ksoftirqd/0]
root	5	0.0	0.0	0	0	?	S<	sep10	0:00	[kworker/0:0H]
root	7	0.0	0.0	0	0	?	S	sep10	0:05	[rcu_sched]
root	8	0.0	0.0	0	0	?	S	sep10	0:00	[rcu_bh]

FIELD	DESCRIPTION
F	PROCESS FLAGS: 　1　Bifurcated but not executed 　4　It has root privileges.
USER	Name of the user who launched the process.
UID	User ID.
PID	ID of the parent process
PPID	ID of the parent process.
PGID	Group ID of a process.
%CPU	Percentage of CPU usage by this process.
%MEM	Percentage of memory occupation by the process.
PRI	Priority of the process. This is the counter field of the task structure. It is the time in HZ of the possible time slice of the process.
NI	Nice, nice value (priority) of the process, a positive number means less processor time and negative time longer (-19 to 19), higher number lower priority.
VSZ	Size of the virtual memory of the process in Kb.
RSS	Size of the resident party; of the physical memory used in Kb.
WCHAN	For the processes that await or asleep, it lists the expected event.
STAT	State of the process: 　R　Executable. (runnable). 　D　Lethargy uninterrupted. (uninterruptible sleep). 　S　Suspended. (sleeping). 　s　It is the leading process of the session. 　T　Stopped, stopped or traced. 　Z　Zombie. 　N　Has a lower priority than normal (indicated in the NI field). 　<　Has a higher priority than normal. 　W　if the process does not have resident pages.
TTY	Name of the terminal to which the process is associated, if there is no terminal then a '?' Appears.
START TIME	Running time.
COMMAND	Name of the command/command/file in execution.

STEP 3: Visualize the states of the running processes in "real" time.

Monitor the processes at runtime.

```
top
ctop
mtop
atop                (broader download application)
```

a)   Help, monitoring application orders.

```
top --help
```

b)   Default display.

```
top
```

```
top - 20:25:56 up  9:12,  3 users,  load average: 0,03, 0,03, 0,05
Tareas: 154 total,   1 ejecutar,  153 hibernar,   0 detener,   0 zombie
%Cpu(s):  4,0 usuario,  0,5 sist,  0,0 adecuado, 95,5 inact,  0,0 en espera,  0,0 hardw int,  0,0 softw int,  0,0 robar tiempo
KiB Mem:   1025940 total,   870288 used,   155652 free,    99296 buffers
KiB Swap:  1046524 total,      440 used,  1046084 free.   374116 cached Mem

  PID USUARIO   PR  NI    VIRT    RES    SHR S  %CPU %MEM     HORA+ ORDEN
 1514 alumno    20   0  321172  87720  35084 S   9,3  8,6  54:24.51 compiz
 1000 root      20   0  126372  26808   8500 S   0,7  2,6   3:52.27 Xorg
    3 root      20   0       0      0      0 S   0,3  0,0   0:07.44 ksoftirqd/0
 2190 root      20   0       0      0      0 S   0,3  0,0   0:06.97 kworker/u4:0
 2229 alumno    20   0   11264   1860   1076 S   0,3  0,2   0:01.22 sshd
    1 root      20   0    4584   2564   1444 S   0,0  0,2   0:01.56 init
    2 root      20   0       0      0      0 S   0,0  0,0   0:00.00 kthreadd
    4 root      20   0       0      0      0 S   0,0  0,0   0:00.00 kworker/0:0
    5 root       0 -20       0      0      0 S   0,0  0,0   0:00.00 kworker/0:0H
    7 root      20   0       0      0      0 S   0,0  0,0   0:01.06 rcu_sched
    8 root      20   0       0      0      0 S   0,0  0,0   0:00.00 rcu_bh
    9 root      rt   0       0      0      0 S   0,0  0,0   0:00.40 migration/0
   10 root      rt   0       0      0      0 S   0,0  0,0   0:01.36 watchdog/0
   11 root      rt   0       0      0      0 S   0,0  0,0   0:01.12 watchdog/1
```

c) Kill a process in another console.
   CTRL+ALT+F2
   (Access with root, visualize the processes and kill a process then change the console)
   ps
   kill -9 numero_proceso

d) Visualize the processes.
   ps
   kill -9 1234
   kill SIGKILL 1234
   Launch a process of sleeping in the background.
   sleep 20 &
   Launch a process in the background, background.
   bg ls -l
   ls -l > /dev/null &
   sleep 100 >/dev/null &

e) Visualize processes.
   ps aux

SIGINT	2	Term	Interrupt from keyboard
SIGQUIT	3	Core	Quit from keyboard
SIGILL	4	Core	Illegal Instruction
SIGABRT	6	Core	Abort signal from abort(3)
SIGFPE	8	Core	Floating point exception
SIGKILL	9	Term	Kill signal
SIGSEGV	11	Core	Invalid memory reference
SIGPIPE	13	Term	Broken pipe: write to pipe with no readers
SIGALRM	14	Term	Timer signal from alarm(2)
SIGTERM	15	Term	Termination signal
SIGUSR1	30,10,16	Term	User-defined signal 1
SIGUSR2	31,12,17	Term	User-defined signal 2
SIGCHLD	20,17,18	Ign	Child stopped or terminated
SIGCONT	19,18,25		Continue if stopped
SIGSTOP	17,19,23	Stop	Stop process
SIGTSTP	18,20,24	Stop	Stop typed at tty
SIGTTIN	21,21,26	Stop	tty input for background process

STEP 4: Kill a process.

The kill command is used to stop processes in the background and also, force to kill a process by sending a signal.
   kill
   killall
   pkill

a) Help.
   kill --help
   kill SIGNAL PID
   kill -numero_señal   PID

b) Kill process (-9) process number 1980.
   kill -9 1980
   ps aux
   jobs

kill [-s] [-l] %pid	
-s	Specify the signal to send. The signal can be a signal name or a number. Process or work identifier.
-l	Write all the signal values supported by the implementation, if no operand is given.
-pid	Process or work identifier.
-9	Force the kill of a process.

c) Maintain a process.
   kill SIGKILL 1001
   kill -9   1001
   End a process.
   kill -15   1002
   kill SIGTERM 1002

d) Help.
   killall --help

e) List all the signals.
   killall -l
   HUP INT QUIT ILL TRAP ABRT IOT BUS FPE KILL USR1 SEGV USR2 PIPE ALRM TERM STKFLT CHLD
   CONT STOP TSTP TTIN TTOU URG XCPU XFSZ VTALRM PROF WINCH IO PWR SYS UNUSED

f) Send a signal to a process.
    killall -s   SIGKILL mio
g) Using the name of a process. The signal to send by default is SIGTERM.
    killall ejemplo001
    killall mysqld
h) Kill a process that does not respond by sending the number of a signal.
    killall 9 mio
i) Help.
    pkill --help
j) Deleted by default.
    pkill ejemplo02
    pkill mysqld
k) Kill a process by selecting the window with the mouse, transformed into a skull.
    xkill

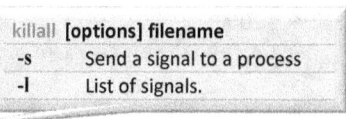

killall [options] filename	
-s	Send a signal to a process
-l	List of signals.

> apt-get install x11-utils
> and it works with htop
> apt-get install htop

STEP 5: Identify / kill the process that is accessing the file.

Show that it processes using so-called files, sockets, or file systems. Identify the processes that make use of a devices.
    fuser

a) Help.
    fuser --help
b) Default. PID shows us the process that shows that my own user is using that directory.
    fuser .
c) List the name of the available signals.
    fuser -l
    HUP INT QUIT ILL TRAP ABRT IOT BUS FPE
    KILL USR1 SEGV USR2 PIPE ALRM TERM STKFLT
    CHLD CONT STOP TSTP TTIN TTOU URG XCPU
    XFSZ VTALRM PROF WINCH IO PWR SYS UNUSED
d) Show unused files.
    root@profesor:/home/alumno# fuser -m /run
    /run: 1   191m   221
    502   511   514   531   532   546   581
    1032   1038 1266 1577 1812 2141 3364
    3422   3482
e) If we execute it on a socket. He will give us this valuable information. As for example in port 80 that starts.
    fuser -v -n tcp 80 webs
    root@profesor:/home/alumno# fuser -v -4
    tcp 80
    El nombre especificado tcp no existe.
    El nombre especificado 80 no existe.
f) Killing selective processes interactively.
    fuser -v -i -k webs

fuser [-fMuvw] [-a|-s] [-4|-6] [-c|-m|-n SPACE] [-k [-i] [-SIGNAL]] NAME	
OPTION	DESCRIPTION
-a	Show unused files too.
-i	Ask before killing (ignored without -k).
-k	Kill processes called file access.
-l	List of names of available signals.
-m	Count display all processes using file systems with name or block device
-M	Fulfill request only if NAME is a mount point.
-n	Search the namespace in this namespace (file, udp or tcp).
-s	Silent operation
-signal	Sending this signal is the place of SIGKILL
-u	ID of the screen user.
-v	Detailed output
-w	Kill only the processes with write access.
-V	Information on the screen version.
-4	Search only in IPv4 sockets
-6	Search only in IPv6 sockets
-	Restart options

STEP 6: Launch a stop process.

    sleep
a) Help.
    sleep --help
b) Assign a default value.
    sleep tiempo
    sleep 1000

bg [options] [process]	
-l	Reports the identifier of the process group and the work folder of the operations.
-p	Reports only the identifier of the process group of operations.
-x	Replaces any job_id found in the command or arguments with the corresponding process group identifier, then executes the command giving it arguments.
job	Specifies the process that wants to run in the background.

STEP 7: Launch a process in the background.

The execution in the foreground is the normal execution, that is, the interpreter does not admit another command until the execution of the current process has finished.

In a terminal only the execution of a single process in the foreground is allowed.

The interpreter allows to execute more than one process in the background.

Background, is to run a process in the background, there are different ways to launch a process in the background.

    bg orden
    orden &

    CTRL+Z
a) Help bg.
    bg --help
b) Launch two processes in the background.
    bg sleep 1000
    sleep 1000 &

c) Stop a running process, in the foreground, pass it to the stopped state.

> sleep 2000
>
> press ^ Z and the process goes to a stopped state

> *# jobs*
>
> *....*
>
> *[5]+ Detenido   sleep 2000 &*

We pass it to the background once the process has stopped, with bg plus% and the number of the job.

> bg   %5

The process [5] that is stopped happens to run in the background.

STEP 8: Visualize the processes in the background.

> jobs

a) Help.

> jobs --help

b) Consult running processes in the background.

> jobs
>
> **[1]+ Ejecutando   sleep 2999 &**

jobs [options]	
-l	Reports the identifier of the process group and the work folder of the operations.
-n	Displays only jobs that have been stopped or closed since the last notification. since the last notification.
-p	Displays only the process identifier for the process group leaders of the selected jobs.

STEP 9: Pass a process from the background to the foreground.

> fg

a) Help.

> fg --help

fg [specifies process]

b) Foreground.

> fg   (jobs)

c) First visualize the running processes in the background, identify the number [2] of the process that you want to execute in the foreground.

> jobs
>
> fg 2

STEP 10: The user execution levels.

The execution levels are considered between the following values: -20 ... .0 ... 19 execution levels -20 has higher priority of execution 19 lower priority, it is normal to launch a process with priority 0, or higher (0 .. 19), the rest of the levels, by default, are controlled by the process scheduler or Shellduler.

> nice   visualize
>
> renice   change / modify

a) Help.

> nice --help
>
> renice --help

b) Display the level of execution that a user has.

> nice

c) Assign the execution level of this user's processes, the default value for a user is 10.

> nice -n 5

renice [-n] <priority> [-p] <pid> [<pid> ...]
renice [-n] <priority> -g <pgrp> [<pgrp> ...]
renice [-n] <priority> -u <user> [<user> ...]

OPTION	DESCRIPTION
-g	Interpret as ID process group.
-n	Set the minimum increment value.
-p	Force to be interpreted as process ID.
-u <name|id>	Interpret as user name or user ID.

d) Reassign the priority level of execution to a process.

> renice -n -2 -p 1834

e) Reassign the level of execution priority to a user.

> renice -n 4 -u   alumno3

f) Reassign the priority level of execution to a group.

> renice -n -2 -g   smr1

STEP 11: Launch a process that does not end even if you restart the computer.

> nohup

a) Help.

> Nohup help

b) Launch a process and endure even if the system is restarted.

> nohup sleep 1000
>
> nohup

STEP 12: Stop a running process.

> stop

a) Help.

> stop   --help

b) Stop a process for your PID, for your identification of the process. (ex .: PID = 4587).

> stop 4587

STEP 13: Change to the execution level system.

The level of execution argument should be one of the multiuser execution levels **2-5**, **0** to stop the system, **6** to restart the system or **1** to let the system down in single user mode.

Normally the tool would be used to stop or restart the system, or to reduce it as a single user.

The level of execution can also be **S** or **s** that the system is placed directly in the single-user mode without having to stop the processes in the first place, it is not usually what is desired.

telinit

a) Help.

   telinit --help

b) Switch to mode 5.

   telinit   5

c) Restart the system ("init 6").

   telinit 6

d) Change in user mode.

   telinit S

   telinit s

e) Change at the execution level stop device.

   init 0

f) Change to the computer restart execution level.

   init 6

g) Restart the computer at the single-user level.

   init 1

h) Multiuser execution level at the moment, for many versions like Ubuntu the level that it has is 5.

   init 5

i) Tell init that it should examine the file /etc/inittab

   telinit q

   telinit Q

j) Perform the action after a number of seconds in: 30 sec.

   telinit -t 30   Q

init, telinit	
Syntax:	init [-a] [-s] [b] [-z xxx] [0123456Ss] telinit [-t seg] [0123456sSQqabcUu]
0, 1, 2, 3, 4, 5 o 6	Levels init that change to the specified execution level
un , b , c	Tell init to process only those entries in /etc/inittab that have the execution level of a, b, or c.
Q o q	Communicate to init to reexamine the /etc/inittab file.
S o s	Tell init to change to single user mode.
U o u	It communicates to init to run again (preserving the state). Do not re-examine the /etc/inittab file. Execution level must be one of S, s, 1, 2, 3, 4, or 5, otherwise the request would be ignored in silence.

Levels runlevel
Runlevel 0 is to stop
Runlevel 1 is single user
Runlevel 2-5 are multi-user
Runlevel 4: As the runlevel 3, but it is not usually used
Runlevel 6 is restart

STEP 14: Contents of the file /etc/inittab.

When starting the system or changing the execution levels with the init or shutdown command, the init daemon initiates the processes by reading the information in the /etc/inittab file. This file defines these important points for the init process:

- That the init process will restart.
- Which processes should be started, monitored and restarted if they are completed.
- What actions should be taken when the system enters a new level of execution.

Each entry in the /etc/inittab file has the following fields:

   id : rstate : action   : process

Descriptions of fields for the inittab file.

FIELD	DESCRIPTION
id	It is a unique identifier for the entry.
rstate	Shows the execution levels to which this entry applies.
action	Identify the way in which the process that is specified in the process field will be executed. Possible values include: sysinit, boot, bootwait, wait and respawn.
process	Define the command or script to execute.

```
[root@localhost ~]# cat /etc/inittab
# inittab is no longer used when using systemd.
#
# ADDING CONFIGURATION HERE WILL HAVE NO EFFECT ON YOUR SYSTEM.
#
# Ctrl-Alt-Delete is handled by /usr/lib/systemd/system/ctrl-alt-del.target
#
# systemd uses 'targets' instead of runlevels. By default, there are two main targets:
#
# multi-user.target: analogous to runlevel 3
# graphical.target: analogous to runlevel 5
#
# To view current default target, run:
# systemctl get-default
#
# To set a default target, run:
# systemctl set-default TARGET.target
#
```

a) Modify the startup mode
   nano /etc/inittab
   # change runlevel boot mode
   id:5:initdefault:

ANNEXES

- *Summary of commands and user management files in Linux.*

- *BiblioWeb.*

- *Compilation of some of the most used Linux commands.*

- *Acronyms.*

Summary of commands and user management files in Linux.

DESCRIPTION:

There are several other commands that are used very little in user administration, which however allow you to manage even more in detail your Linux users. Some of these commands allow you to do the same as previously seen commands, only in other ways, and others such as 'chpasswd' and 'newusers' are very useful and practical when you register multiple users.

Below I present a summary of the commands and files seen in this tutorial plus others that a little research.

User management and control commands

adduser	See useradd
chage	Allows you to change or set parameters of the password control dates.
chpasswd	Update or set passwords in batch mode, multiple users at the same time. (used together with newusers)
id	It shows the identity of the user (UID) and the groups to which it belongs.
gpasswd	Manage group passwords (/etc/group and /etc/gshadow).
groupadd	Add groups to the system (/etc/group).
groupdel	Remove groups from the system.
groupmod	Modify system groups.
groups	Shows the groups to which the user belongs.
newusers	Update or create users in batch mode, multiple users at the same time. (used together chpasswd)
pwconv	Set the shadow protection (/etc/shadow) to the /etc/passwd file.
pwunconv	Remove the shadow protection (/etc/shadow) from the /etc/passwd file.
useradd	Add users to the system (/etc/passwd).
userdel	Remove users from the system.
usermod	Modify users.

Administration and user control files

.bash_logout	It is executed when the user leaves the session.
.bash_profile	It is executed when the user starts the session.
.bashrc	It is executed when the user starts the session.
/etc/group	Users and their groups.
/etc/gshadow	Encrypted passwords of the groups.
/etc/login.defs	Variables that control the aspects of user creation.
/etc/passwd	System users
/etc/shadow	Encrypted passwords and date control of system users.
/etc/inittab	This file defines these important points for the init process.
/etc/fstab	List of available disks and partitions, how to mount each device and which configuration to use.
/etc/mtab	It contains a list of the currently mounted file systems.
/etc/resolv.conf	It contains the IP addresses of the name servers (DNS name resolvers) that will try to translate names to addresses for any available node in the network.
/etc/newtworks/interfaces	Give your network card an IP address (or use dhcp), set routing information, configure IP masquerading, set the default routes.
/etc/hostname	Stores the main name of a computer.
/etc/apt/sources.list	They list the available "sources" or "repositories" of software packages to be: updated, installed, removed, searched, subject to comparison of versions, etc.
/proc/meminfo	It contains the information of the RAM memory.
/proc/cpuinfo	It shows all the information of our processor and in the case of being dual-core, it appears as if they were two.
/etc/inittab	Change the execution levels.
/var/run/utmp	Registers users connected to the system in real time.
/var/log/wtmp	It contains the structures of certain log files as lastlog.
/etc/shells	List of interpreters of supported commands.
/etc/tercamp	The capacity database of the terminal.
/etc/motd	Contains the message of the day, which is issued automatically after starting a successful session on the ftp server.

BiblioWeb

Web References

SISTEMA OPERATIVO /CONCEPTO	URLs
Android for PC	http://www.android-x86.org/download
	http://developer.android.com/index.html
Help	http://manpages.ubuntu.com/manpages/dapper/es/
	http://es.hscripts.com/tutoriales/linux-commands/head.html
	http://cmaverick.wordpress.com/comandos-linux/
	http://es.kioskea.net/faq/3435-linux-comandos-para-monitorear-el-sistema
	http://www.alcancelibre.org
	http://www.elblogderigo.info/2013/12/12/muchos-tips-trucos-y-mas-para-linux/
	http://www.bdat.net/shell/book1.html
CentOS	http://www.centos.org/download/
Graphics Drivers	https://wiki.archlinux.org/index.php/Xorg_(Espa%C3%B1ol)
Debian	https://www.debian.org/distrib/
	http://www.debian.org/doc/manuals/apt-howto/index.es.html
Edubuntu	https://edubuntu.org/download
Fedora	http://fedoraproject.org/get-fedora
Lubuntu	https://help.ubuntu.com/community/Lubuntu/GetLubuntu/
Mint	http://www.linuxmint.com/download.php
Putty	http://www.putty.org/
Repositories for Slackware	http://connie.slackware.com/~alien/slackbuilds/ Alien
	http://rlworkman.net/pkgs/ For Robby
	http://www.slacky.eu/ For the Italian community
Slackware	http://mirrors.slackware.com/slackware/
Suse	https://download.suse.com/index.jsp
Ubuntu	http://www.ubuntu.com/download/desktop
Help Script	http://persoal.citius.usc.es/tf.pena/ASR/Tema_2html/
	https://www.gnu.org/software/bash/manual/html_node/Shell-Builtin-Commands.html
File Systems	http://somebooks.es/capitulo-3-estructura-del-sistema-operativo/7/
	https://wiki.archlinux.org/index.php/fstab

Compilation of some of the most used LINUX commands.

A

addgroup	Shell Scrip that allows you to add a group.
adduser	Shell Script to create a user.
alias	In some cases, commands that are difficult to remember or that are too long are often used, but in UNIX there is the possibility of giving an alternative name to a command so that each time you want to execute it, only use the alternative name .
apt-cache search (text)	It shows a list of all the packages and a brief description related to the text that we have searched for.
apt-get dist-upgrade	Additional function of the previous option that modifies the dependencies by the new versions of the packages.
apt-get install (package)	Install packages
apt-get remove (package)	Delete packages. With the -purge option we also delete the configuration of the installed packages.
apt-get update	Update the list of available packages to install.
apt-get upgrade	Install the new versions of the different packages available.
at	Perform a scheduled task only once.
atop	Monitor the execution of processes.

B

bash, sh	There are several shells for Unix, Korn-Shell (ksh), Bourne-Shell (sh), C-Shell (csh), bash.
bg	Send a process to the background.
biosdecode	It analyzes the BIOS memory and prints the information about all the structures (or points of entry).

C

cal, ncal	Show the calendar
calendar	Displays the ephemeris of a calendar date
cat	Shows the content of the file on the screen continuously, the prompt will return once the contents of the entire file have been displayed. It allows you to concatenate one or more text files.
cd	Change directory
chattr	Change attributes of a file.
chgrp	Change the group to which the file belongs.
chmod	Used to change protection or access permissions to files. r: read w: write x: execute +: add permissions -: remove permissions u: user g: user group or: others.
chown	Change the owner of a file.
chroot	It allows us to change the root directory.
clear	Clean the screen, and place the prompt at the beginning of it.
cmp, diff ,comm	It allows the comparison of two files, line by line. It is used to compare data files.
cp	Copy files to the indicated directory.
crontab	Perform a scheduled task on a regular basis. It allows to monitor processes, it offers a dynamic view of the activity of the processor in real time.
ctop	Its main use is to show a column of a certain output. The -d option is followed by the delimiter of the fields and the -f option is followed by the field number to be displayed.
cut	The "delimiter" by default is the tabulator, we change it with the -d option. It has some other useful options.

D

date	Returns the day, date, time (with minutes and seconds) and year.
df	Shows the mounted file systems.
dmesg	Displays kernel messages during system startup.
dmidecode	It allows to know thoroughly the hardware of our equipment, as it is described in the BIOS of the system according to the SMBIOS / DMI standard SMBIOS; which means "System Management BIOS" and DMI stands for "Desktop Management Interface"
dpkg-reconfigure (package)	Reconfigure an already installed package.
du	It serves to see what each directory occupies me within the directory in which I am and the total size.

E

echo	Displays a message on the screen.
eject	By using this command the ejection of the CD unit will be achieved, as long as it is not in use.
env	To see the global variables.
exit	Close the windows or the remote connections established or the shells open. Before leaving it is advisable to eliminate all jobs or processes from the workstation.
egrep	Search and find in one or more files lines that match the string or word given.

F

fdisk,cfdisk	Visualize and establish partitions, file system types and everything related to the MBR.
fg	Send a process to the foreground.
file	Determine the type of the indicated file (s).

find	Find the files that satisfy the condition in the indicated directory.
finger	It allows to find information about a user.
free	It shows information about the state of the system memory, both the swap and the physical memory. It also shows the buffer used by the kernel.
fgrep	Search one or more files for lines that match the string or word given. fgrep is faster than the grep search, but less flexible: you can only find text, not regular expressions.
fsck	To check if there are errors in our hard drive.

G

gdisk, cgdisk	Visualize and establish partitions, file system types and everything related to MBR, GPT and others.
gpasswd	Facilitates the task of managing a group of users.
grep	Its functionality is to write in standard output those lines that match a pattern. Look for patterns in files.
gzip	Compress single file using the .gz extension.
groupadd	Create a new group of users.
groupdel	Remove a group of users.
groupmod	Modify a group of users.
groups	Shows the groups to which the user belongs.

H

halt	It allows to turn off, restart the computer and in turn synchronize.
head	Shows the first lines of a file.
history	Lists the most recent commands that have been entered in the window. It is used to repeat commands already typed, with the command!
hostid	Displays the numeric identifier of the current host in hexadecimal.
hostname	Displays or sets the name of the machine or machine.

I

id	ID number of a user.
ifconfig	Obtain information about the network configuration.
info, infotext	Displays the information about the commands on a navigable screen equivalent to man.
init	Change the RUNLEVEL execution level.

J

job	Lists the processes that are running in the background.

K

kill	It allows to interact with any process sending signals. Kill (pid) ends a process and Kill -9 (pid) forces to finish a process in case the previous option fails.
killall	Send a signal to all processes with the same name.

L

last	This command allows you to see the last connections that have taken place.
less	Show the file in the same way as you would, but you can return to the previous page by pressing the "u" or "b" keys.
ln	It serves to create links to files, that is, create a file that points to another. It can be symbolic if we use -s or hard link.
lock	It allows to block the terminal, for it asks for a password, twice.
locate	Locate files by consulting the updatedb database.
logout	The sessions end with the logout command.
logname	Shows the current login
last	Lists the last users connected to the system.
lastb	It shows the failed accesses of the connection (s) of a user.
lastlog	Show the last connection time of the system accounts. The access information is read from the /var/log/lastlog file.
less	Visualize the files by pages and allows the advance and the backward movement. Allows access to filters.
ls	Lists the files and directories within the working directory.
lsattr	See attributes of a file.
lshal	Display of the elements in the database of the HAL device.
lsmod	Shows the modules loaded in memory.
lsoff	Lists the files opened in the system.
lspci	List all pci-type components (Peripheral Component Interconnec).
lsusb	Displays all connected USB devices.
lwclock	Utility to access the Hardware clock.

M

man	It offers information about the commands or topics of the UNIX system, as well as existing programs and libraries.
mesg	Activate or cancel the broadcast of messages with write.
mkdir	Create a new directory.
mknod	Create special files of character / block devices.
mv	This command is used to rename a set.
more	Show the file on the screen. Pressing enter, line by line is displayed. Pressing the spacebar, screen by screen. If

	you want to exit, press q.
mount	In Linux there are no A: or C: units but all devices "hang" from the root / directory. To access a disk, you must first mount it, that is, assign it a place in the system's directory tree.
mtop	It allows to monitor the execution of the processes in real time, external application.
mv	Move files or subdirectories from one directory to another, or change the name of the file or directory.

N

nano	Text editor in the order line, editor similar to WordPerft (same peak).
newgrp	It allows to change from the current user to another group (we need to know the password).
nice	It allows to change the priority of a process in our system.
nohup	It allows a process to continue its execution when the computer is restarted, if during the execution a system crash occurred, it will return to the execution point that remained before the crash.

O

No order is dealt with this letter.

P

passwd	It is used to set the password to a user.
paste	It unites two files laterally.
pico	Text editor on the order line equal to nano.
ping	The ping command is usually used to test aspects of the network, such as checking that a system is on and connected; this is achieved by sending ICMP packets to said machine. The ping is useful for verifying TCP / IP installations. This program tells us the exact time it takes for the data packets to go and return through the network from our PC to a specific remote server.
pg	It allows to visualize files of flat text in scroll, with displacement of edition, identical to more.
pmap	Memory map report of a process (s).
poweroff	Turn off the computer.
ps	Displays information about active processes. Without options, it shows the process number, terminal, cumulative execution time and the name of the command.
pstree	Shows a process tree.
pwck	Verify the integrity of password files.
pwd	Shows the current working directory.

Q

No command that starts with this letter has yet been treated.

R

reboot	Restart the system is called when the system is not at levels 0 or 6, under normal conditions.
reset	If we observe that we write on the screen and the text does not appear but when pressing enter is really being written, or that the colors or texts of the console are corrupted, it may be that some application in text mode has ended abruptly, not restoring the standard values of the console when leaving. With this we force some default values, regenerating the screen.
rlogin	They connect a local host with a remote host.
rm	Remove or delete a file.
rmdir	Remove the indicated directory, which must be empty.
rmmod	Memory discharge a module, but only if it is not being used.
renice	Redefines the user's priority.
route	The route command is used to display and modify the routing table.

S

sar	Statistical page display.
set	To see the environment variables.
slapt-get	It is an APT-based system for handling packages in the Slackware GNU / Linux distribution.
sleep	Launch a process for a time in thousandths of seconds.
shutdown	Automatic shutdown in Linux.
sort	It shows the contents of a file, but showing its lines in alphabetical order.
ssh (Secure Shell Client)	It is a program to connect to a remote machine and run programs on it. Used to replace the rlogin and rsh, it also provides greater security in the communication between two hosts. The ssh connects to the indicated host, where the user enters their identification (login and password) on the remote machine, which performs a user authentication.
startx	Start the graphical environment (X server).
stop	For a process.
stty	Displays the tty terminals connected in series.
su	With this command we access the system as root.
sum	Display the checksum of a file.
symlink	Manipulation symbolic link.
sync	Synchronize the data on the disk with the memory.

T

tac	It allows to visualize the content of a plain text file in reverse format, from the last line to the first one. It is the inverse to cat.
tail	This command is used to examine the last lines of a file.

tar	Compress files and directories using the .tar extension.
telnet	Connect the local host to a remote host, using the TELNET interface.
top	It shows the processes that are executed at that moment, knowing the resources that are being consumed (Memory, CPU, ...). It is a mix of the uptime, free and ps command.
touch	Create an empty file.
tee	It allows to redirect to multiple files, use with filters.
Telinit, init	Initialization of process control.
tty	It allows to visualize the open consoles in tty or PTS0.

U

umask	Set the permission mask. The permissions with which the directories and files are created by default.
umount	Unmount mounted units. You do not need to specify the device only the mount point.
unalias	Delete an alias
uname	Displays the system information.
uniq	This command reads an input file and compares the adjacent lines by writing only one copy of the lines to the output. The second and subsequent copies of the repeated adjacent entry lines will not be written. Repeated lines will not be detected unless they are adjacent. If no input file is specified, the standard input is assumed.
unset	It sets the value of the variables to zero, if they are queried by them after setting them to zero, a null string will be shown (a blank line).
uptime	It tells us how long the machine has been running.
useradd	Create a new user
userdel	Delete existing user
usermod	Modify an existing user.
users	Show the connected users.

V

vi	It allows editing a file in the current working directory. It is one of the most used text editors in LINUX and formerly in UNIX.
view	It is similar to vi, only it does not allow to save modifications in the file, it is to read the contents of the file.

W

wathis	Brief description of a command.
wc	Count the characters, words and lines of the text file.
whereis	Returns the location of the specified file, if it exists.
which	Find the location of the command in the Path directories (whereis).
who, w	List of those connected to the server, with username, connection time and the remote computer from which it is connected.
whoami	Enter your username on the screen.
write	Send a message to another user's terminal.

X

xinit, startx	Start or launch the X Windows server.

Y

yes	Write the 'y' character or the message indefinitely.
yum	The Yellowdog Updater, Modified (mmm) is an open source command-line package utility for Linux operating systems using the RPM package manager.

Z

zcat	Display the contents of a text file, compressed with zg format.
zdiff, zcmp	Compare compressed files.
zmore, zless	Visualize the content of text files, it is in zg format.

Acronyms

AMD	Advanced Micro Devices, Inc. Company dedicated to the development of microprocessors and other integrated circuits such as video cards
AMD-V	AMD Virtualization.
API	Application Programming Interface.
APIC	Advanced Programmable Interrupt Controller.
APM	Advanced Power Management.
APT	Advanced Packaging Tool.
ARM	*Advanced RISC Machine.*
BIOS	Basic Input/Output System.
BSD	Berkeley Software Distribution.
CIFS	Common Internet File System.
CPU	Central Processing Unit.
EFS	Encrypting File System, sistema de ficheros encriptados.
ext3	Third extended filesystem o tercer sistema de archivos extendido.
ext4	Fourth extended filesystem o «cuarto sistema de archivos extendido.
FAT	File Allocation Table.
FTP	File Transfer Protocol.
GID	Group Identifier.
GNOME	GNU Network Object Model Environment.
GPG	GNU Privacy Guard (GTnuPG o GPG) es una herramienta de cifrado y firmas.
GRUB	GNU GRand Unified Bootloader, es un gestor de arranque múltiple.
GUI	Globally Unique Identifier.
HTTP	The Hypertext Transfer Protocol.
IDE	Integrated Device Electronics.
IMEI	International Mobile Equipment Identity.
ISO	International Organization for Standardization.
KDE	K Desktop Environment o Entorno de Escritorio K.
LBA	Logical Block Addressing.
LILO	Linux Loader.
LISP	LISt Processing.
LVM	Logical Volume Manager.
MFT	Master File Table.
MS-DOS	MicroSoft Disk Operating System.
NTFS	New Technology File System.
NX	*No eXecute*, Bit de Procesador, puede ser DX. Ayuda al procesador a proteger al equipo contra ataques de software malintencionado.
PAE	*Physical Address Extension.* Physical address extension. Allows 32-bit processors to access more than 4 GB of physical memory in compatible versions of Windows and is a prerequisite for NX.
RAID	Redundant Array of Independent Disks.
RAM	Random Access Memory.
SAMBA	Server Message Block Protocol.
SATA	Serial Advanced Technology Attachment.
SGL	Es la base de la tecnología de Google para gráficos en móviles.
SSH	Secure Shell.
SMB	*Server message block.*
SO	Sistema Operativo (OS Operating System).
SPARC	Scholarly Publishing and Academic Resources Coalition.
SQL	Structured Query Language.
SSL	Secure Sockets Layer.
UDI	Uniform Driver Interface.
UID	User ID.
VT-X	Enable Intel Virtualization Technology.
WOL	Wake On Lan. It is a standard network of Ethernet computers that allows you to turn them on remotely (they are off).